THE SECOND CREATION

THE SECOND CREATION

THE AGE OF BIOLOGICAL CONTROL BY THE SCIENTISTS WHO CLONED DOLLY

Ian Wilmut, Keith Campbell
and Colin Tudge

HEADLINE

First published in 2000
by HEADLINE BOOK PUBLISHING

10 9 8 7 6 5 4 3 2 1

Appendix reprinted by permission from *Nature*, Vol. 385, pp.811–13.
Copyright © 1997 Macmillan Magazines Ltd

British Library Cataloguing in Publication Data

Wilmut, Ian
   The second creation : the age of biological control
   1. Cloning
   I. Title II. Campbell, Keith III. Tudge, Colin, 1943–
   660.6'5

Hardback ISBN 0 7472 2135 9
Trade paperback ISBN 0 7472 7530 0

Photographs supplied by Design and Print
Services, Roslin Institute
Typeset by Palimpsest Book Production Limited,
Polmont, Stirlingshire
Printed and bound by
Mackays of Chatham plc, Chatham, Kent

HEADLINE BOOK PUBLISHING
A division of the Hodder Headline Group
338 Euston Road
London NW1 3BH

www.headline.co.uk
www.hodderheadline.com

CONTENTS

PART IV: *The Age of Biological Control*

The Place and the People

I first met Dolly in March 1997, a few weeks after Ian Wilmut, Keith Campbell and their colleagues announced her existence in the scientific journal *Nature*. She was a sweet creature, much cosseted since birth and extraordinarily tame. She danced round her pen at Roslin Institute as eagerly and vociferously as a spaniel, leaping into her feed trough for a better view, sociable even by the standards of sheep. Roslin's other famous clones, Megan and Morag, who were born a year before her in 1995, had been drafted in to keep her company.

Many dire predictions followed Dolly's birth. Some were obvious fantasy – or at least, we may hope the fantasy was obvious – as in one American newspaper which announced that she was a carnivore and ate her flock-mates. Others were more serious. Dolly had been cloned from a cell from a six-year-old ewe – so were her own cells merely as old as she was, or should we add six years to their biological age? Six years is a lot, in the life of a sheep. In recent years, too, biologists have discovered a phenomenon called 'genomic imprinting', which says that a gene inherited from the mother may behave differently from the same gene inherited from the father. All will be explained later. Suffice to say here that some biologists feared that because Dolly had been cloned from a body cell, and not by the fusion of egg and sperm, then genomic imprinting would cause problems. Others suggested, albeit for no very obvious reason, that she would be sterile.

But Dolly seems to have defied the Jeremiahs – at least in ways that seem to matter. True, in the summer of 1999 structures at the end of her chromosomes known as telomeres, which become shorter with age, were more typical of a much older animal; but she herself shows no sign of premature ageing. This certainly does not imply that cloning is now open and shut, and that any animal (or human being!) can now be cloned at will in perfect safety. Dolly represents one success out of several hundred attempts. But, at least in her case, the worst fears seem to be confounded. At eight months she behaved and looked like an eight-month-old ewe and in the summer of 1999 she was a very typical three year old.

Genomic imprinting has caused no obvious problems and she is certainly not sterile. When I last went to see her, in the autumn of 1998, she had her daughter at heel – named 'Bonnie' because, as the vet Tim King commented, 'She is a bonnie wee lamb'. Bonnie's father is David – one of the tiny elite among sheep to be given a name. 'Reporters ask us what the sheep are called,' says Tim King, 'so we just have to make something up. David is a Welsh Mountain ram, so it seemed appropriate.' David grazed moodily outside in the gathering November gloom while Dolly remained in her shed, which by that time she shared with the entire flock of Roslin clones: Megan and Morag (Megan also with a daughter); Taffy and Tweed – males cloned from Welsh Black fetal cells, and very black indeed; a cloned quartet of young rams – Cedric, Cyril, Cecil and Tuppence; and the ewe that should perhaps be the most famous of all – Polly. Polly, born a year after Dolly in 1997, is not only cloned, she is also genetically transformed. She has been fitted with a human gene that codes the protein factor IX, which she secretes in her milk. This potentially has huge therapeutic value. Factor IX is involved in blood clotting and its deficiency causes a form of haemophilia.

I became involved through my restlessly intellectual agent, John Brockman, who saw the report in *Nature* and suggested that I should call Ian Wilmut and propose that we write a book together. Ian agreed – though stressing that his colleague Keith Campbell had made the biggest innovation in the research and so must also be a coauthor. Ian had been one of Roslin's front-men after Dolly was announced, together with the institute's assistant director Harry Griffin, so very few people knew about Keith. I certainly did not. But Keith's role, as we will see, has been vital.

So I have worked at this book over the past two years, rushing up from London to Roslin (near Edinburgh) every few months, for another burst of information and ideas. The exercise has proved far more difficult than I imagined. Notably, the basic science has turned out to be much more complicated than I had foreseen, and the research has far, far greater implications than most commentators seem to have realised. Most, for example, have homed in on the prospect of human cloning, but human cloning is really a side issue – and one that Wilmut and Campbell find distasteful. So this account is correspondingly more intricate than I had anticipated. But the work poses a challenge for everybody: a test case. It is immensely important, with ramifications that will affect all our lives in a dozen different ways. If the concept of democracy means anything, then it should imply that people at large understand the forces at work that change society. So everyone in a democracy *ought* to understand the research at Roslin – not just the cloning itself, but all that flows from it. The challenge is for me, Ian and Keith to convey its complexities and subtleties, and for people at large to take an interest. If people do not get to grips with such research then the new biotechnologies that are beginning to influence our lives so profoundly will forever occupy some esoteric, elevated niche, effectively out of sight. That would be a huge pity, for cultural and for political reasons. On the whole, the world at large deploys science and technology with little finesse – we seem to suffer many of their ill effects, without properly reaping the many possible benefits – and we will never do better unless people, meaning voters, consumers and citizens, have a feel for what is going on.

Roslin's background and a few details of its history are needed to round off the story – who's who, and why things are called what they are. Without this brief outline the grand narrative can be confusing – not least because Roslin has changed its name several times since Ian joined, and once even in Keith's time. In her excellent book *Clone*, which she produced within a few months of Dolly's birth, Gina Kolata remarks that some American commentators in particular have difficulty in understanding exactly what kind of place Roslin *is*. Is it a university department? Is it a commercial company? Why, if its scientists do serious science, do they work mainly on sheep, rather than mice? Some wrote about Roslin as if it was a kind of cow-shed, an accident, quaintly guarded by a large dog (who apparently is called Buster, although I cannot vouch for this).

In truth, Roslin is not a commercial company – though it does have several commercial biotech companies on its campus, one of which (known as PPL) was heavily involved in the creation of Dolly and Polly. Neither is Roslin a university department. In fact, it is a government laboratory, one of a network of institutes throughout Britain that now answer to the Biotechnology and Biological Sciences Research Council (BBSRC) which in turn is one of seven such councils that together form the core and umbrella of Britain's government-supported science and technology. (Others include the Medical Research Council (MRC) and the National Environmental Research Council (NERC).) Roslin also receives a significant proportion of its funding from various ministries including the Ministry of Agriculture, Fisheries and Food (MAFF) and from industry.

Roslin, founded in 1919, has been called 'Roslin Institute' (not '*the* Roslin Institute') only since 1993, but it can trace its origins back to 1911 when the British government set up a commission to try to raise the general level of Britain's agriculture. Out of this commission came agricultural boards, which in 1931 begat the Agricultural Research Council (ARC), which in turn in October 1983 became the Agricultural and Food Research Council (AFRC), which finally, in 1994, was subsumed within the BBSRC.

This succession of overseeing bodies has established many new agricultural laboratories and 'units' throughout Britain but they have also perforce been opportunist – making use of what was there. All the government labs have established links with universities. Roslin (as now it is) can trace some of its roots to, and still maintains strong links with, Edinburgh University; some of the initiatives that have been incorporated into Roslin began there. The overseeing bodies have also reflected the changing nature of science. Thus 'classical' genetics (which in agriculture is of course linked to livestock and crop breeding) was founded by Gregor Mendel in the 1860s but it did not become established as a formal science until the 20th century. Edinburgh University has been a prime centre of genetic studies from earliest times, and these have fed into the modern Roslin.

In the 1950s James Watson and Francis Crick at Cambridge described the three-dimensional structure of DNA (short for '*deoxyribonucleic acid*') and thus founded the modern discipline of 'molecular biology' – essentially the science of DNA and its immediate products. Since DNA is the stuff of which genes are made, molecular biology clearly complements classical genetics and the traditional, more venerable disciplines of physiology and biochemistry. The traditional disciplines have to take account of molecular biology, and embrace what it has to offer. Molecular biology in particular, too, cuts across the conventional divide between agriculture and medicine: thus Roslin has worked with farm animals, but its cloning research has immediate relevance to medicine (*vide* Polly), as you will see. Finally, outstanding scientists do not grow on trees and the succession of ruling bodies has always tried to build laboratories around outstanding individuals – like the great geneticist C. H. Waddington, at Edinburgh. All these factors – the way that science has changed, and the need to build on what already exists and on the special talents of individuals – have complicated the history of government-backed research, and of the institutes that emerge from it.

However, from the 1970s onwards, scientific research throughout the western world has been beleaguered by deeper issues. Until the 1970s scientists who received salaries directly from governments or from universities tended to keep their distance from industry. They felt that it would compromise their calling to apply their work for mere gain. There was haughtiness in this, but also a kind of saintliness; thus Cesar Milstein as late as the early 1970s omitted to patent his discovery of monoclonal antibodies, which have transformed all biotechnology, because he felt they should be used simply for the good of humankind. By the 1980s, however, particularly under the governments of Margaret Thatcher and Ronald Reagan, the attitude was reversed. Governments now felt that science *ought* to be seen to pay its way. Where government-based and university scientists in an earlier, more innocent age had been expressly forbidden to profit from their endeavours, they were now encouraged to form industrial links, to create profits, and to feed those profits back into their institutions. For this reason, much research these days is done on 'soft' money (*ad hoc* grants from outsiders). Scientists who once were content to take their salaries and draw their research funds from their institutions now typically spend several working days each month writing applications for grants, often though not exclusively from industry, to finance their work and sometimes their own salaries. About one in five or six of these applications succeeds.

No one finds it easy, however, to draw lines between the kind of research that government should support, and the kind that industry should support; and various kinds of research, including the work that has led to Dolly and the rest, tends to be caught in the middle. Thus at one time in the recent past the government councils encouraged their own scientists to focus on science that was not too 'pure', but would have some pay-off in the foreseeable future; but, later, government said that if the research was obviously liable to be profitable before too long then industry should pay for it. Where stands Dolly on this spectrum of applicability? Overall, too, public spending has been reduced over the past two decades so that many government institutions have been severely squeezed, with huge rafts of redundancies, and some have closed altogether. It was such a drawing in of horns in 1982 that forced Ian to change tack: from research on prenatal mortality (notably in sheep) and into the work that led to Dolly. Overall, then, we may paint grand visions of the scientific future and see very clearly where various lines of research ought logically to lead, but what actually gets done is what people are prepared to pay for, and that, increasingly, is a matter of catching a sponsor's eye. Science certainly progresses, but not so logically as is often supposed.

At the time that Ian joined what is now Roslin, in 1973, as a 'senior scientific officer', it was still called the Animal Breeding Organisation, or ABRO, and was still answerable to the ARC. Later – in 1986 – ABRO was joined with the AFRC's Poultry Research Centre (PRC) to form the Institute of Animal Physiology and Genetics Research (IAPGR). Chickens are still very much in evidence at Roslin, with cut-price eggs a weekly bonus. In fact, though, the Edinburgh branch formed only half of IAPGR; the other half was in Cambridge, so what is now Roslin was properly called the Edinburgh Research Station of the AFRC Institute of Animal Physiology and Genetics Research (the 'ERS of AFRC's IAPGR'). It was in this form when Keith joined in 1991. Mercifully, and perhaps to save ink, it was renamed 'Roslin Institute' in 1993, when it also became independent of Cambridge. Currently Roslin does research on all the major farm animals – on their molecular biology, genetics, embryology and reproduction, development and growth, and animal welfare. It employs 300 people at any one time and, now that the mining industry of the Scottish borders has been shut down, Roslin is one of the largest employers in the area, although behind tourism. Scientists come from all over the world to work at Roslin – including PhD students, 'postdocs' and senior scientists. A backwater it certainly is not. And serious research does not have to be done on mice.

There is one more significant player in the cloning story: the commercial biotech company PPL. Just as ARC was putting the squeeze on ABRO in the early 1980s, ABRO's then-director, John King, was seeking to introduce more molecular biology into the institute – and to this end he wanted more funds, rather than less. King was shortly succeeded by Roger Land, who invited John Bishop in the department of genetics at Edinburgh to be a special consultant. ABRO then appointed its first full-time molecular biologist, Rick Lathe, who worked at the biotechnology laboratory Transgene in Strasbourg, France. Bishop and Lathe came up with the idea of producing therapeutically useful proteins in the milk of livestock, and to this end they established a biotech company known initially as Caledonian Transgenics, which was based in the ABRO building (which in turn was at that time based on the campus at Edinburgh University). Rick Lathe then returned to France and was succeeded by John Clark, who has been a key player ever since and remains at Roslin as head of the division of molecular biology. On to the scene at this point came Alan Colman, then professor of biochemistry at Birmingham University, and Dr Ron James, a chemist by training, who were seeking ways of producing therapeutic proteins in animals – and at that time felt that chicken's eggs might be a good vehicle for the necessary human genes. But at ABRO they found the work on milk already beginning, and in 1987, with venture capital, they converted Caledonian Transgenics into Pharmaceutical Proteins Ltd, which is now known as PPL. PPL is now floated on the stock market, and had an initial value of £120 million. In 1993 PPL merged with TransPharm Inc. (now PPL Inc.) of Blacksburg, Virginia. While Caledonian Transgenics was mutating into PPL, ABRO was undergoing its various transformations to become Roslin Institute. In the 1970s Roslin moved to its present site, and PPL then took up residence on the same site.

PPL and Roslin are separate establishments – Roslin has no direct financial interest in PPL because of the rules that prevailed in the early 1980s – but the two continue to cooperate. Thus PPL provided the cells from which Dolly was made, and some of the cash; and PPL scientists were the principal creators of Polly. Keith went on to work for PPL, although he continued to collaborate with former colleagues at Roslin, and subsequently moved to Nottingham University to become a professor. Just to provide the final complication, in 1998 Roslin founded its own commercial company – as they were not allowed to do at the time that PPL was established! – called Roslin Bio-Med, which is also established on the Roslin site, and to which Ian is a consultant. In May 1999 Roslin Bio-Med merged with Geron, a California company based near San Francisco. Roslin Bio-Med and PPL have somewhat different agenda but there could be areas of competition between the two.

Thus until Keith moved to Nottingham he and Ian continued to work in the same field and on the same campus, often attended the same conferences, lived in the same area, and of course conversed but, says Ian, 'We were always aware of commercial constraints'. All in all, Roslin and PPL are good friends although their relationship overall is very like that of the British and French in World War I, at least as described by A. J. P. Taylor: allies right enough, but sharing delicate information if at all only on a need-to-know basis. All in all, the administration and politics of science can be just as complicated as the science itself, although whether you find it interesting depends on your point of view.

Roslin Institute itself is a long, smart, glassy, 1970s sort of building with steep roofs, and mostly of one storey, not least because the ground beneath is riddled with the tunnels of miners and a heavier structure might work its way down into labyrinths. Now the mines have gone – the slag-heaps and pitheads are cleared away, although the mining villages remain, with fine border names like Penicuik and Bonnyrigg – and the surrounding area is officially depressed, in receipt of European Union support. To the visitor, however, and to locals who still have jobs, the landscape is a delight. The village of Roslin where the Institute stands peaked in the 15th century, when it was reputedly more important than Edinburgh itself. Many a luminary has passed through, including Dr Johnson, who reacted to it less dyspeptically than he generally did to things Scottish, and went so far as to say that he found it quite agreeable. Roslin village has a ruined castle (now a writers' retreat) and a chapel founded by the Knights Templar, deeply carved with semi-pagan images, at once both down to earth and mystical; plough-boys wind their way up the columns and into heaven. The 'Horizon' television team who came to film Dolly took the strivings of those anxious peasants as a symbol for the hopes and hubris of modern technology, and dwelt long and lovingly upon them. But the chapel still runs very pleasant (and impeccably Christian) services. The surrounding countryside is – well, lowland Scottish: lively skies, dark trees on bright green hills, white hotels like chateaux with steep black roofs and pepperpot towers, and flock upon flock of Scottish Blackface sheep. It's a brisk, no-nonsense, working landscape. Very sharp, very refreshing – just the place to practise the traditional pursuits of drinking, walking and staring into space. This kind of country gave rise to the Scottish Enlightenment and Romanticism. David

Hume, Adam Smith, and Walter Scott wandered and contemplated in these hills. Thomas Carlyle was born just down the road in the village of Ecclefechan, where his statue now glares grumpily through the drizzle. All in all it is a place where people have very grand ideas of a very practical nature that change the course of human history. It is absolutely appropriate as the birthplace of mammalian cloning.

One further difficulty we faced in writing this book was technical: the matter of voice. I have put the words down on paper – so should I write in the third person, referring to the two principal players, Wilmut and Campbell, as 'Ian' and 'Keith'? I tried that approach, but it didn't work. This is *their* book, *their* story, told to me in a series of long, taped interviews (supplemented by days in specialist libraries). The only appropriate voice, therefore, was the first person. But Ian and Keith have played different parts. Their careers before they started to work together are pertinent to the story, and even when they were working on the same part of the project together – on Megan and Morag, for example – they did different things. So it was also not always appropriate to say '*We* did this' or '*We* did that'. To avoid switching voices between paragraphs, I have written some chapters from Ian's point of view and some from Keith's, with a few from both. As we will see, it was Keith's inspiration that finally brought success; but Ian conceived the project and is its leader, and it is appropriate that he should provide the overall picture.

Finally, there is a point of style. The story of cloning is extraordinary, and it will change all our lives. It is hard to overstate its implications. Cliché has it that scientists who do extraordinary things must themselves be extraordinary, and indeed somewhat crazed. Doctors Frankenstein, Moreau and Strangelove are among the 'weird sisters' of literature. It is common, in 'popular' books, to write about science and scientists in hushed tones, as if all the participants were somewhat unstable geniuses, liable to pour contempt on the smallest solecism or to fly off into some unfathomable rage. Perhaps there are scientists like that. There are certainly some who think too much of themselves, and build themselves a carapace of eccentricity. But scientists for the most part are not like that at all. They think extraordinary thoughts, and sometimes – this is the privilege of the job – they are able to turn those thoughts into reality. Dolly is extraordinary, yet she has been made manifest.

But it would not be an insult to Ian and Keith – really rather the opposite – to suggest that they are, in the parlance of Middle England, 'ordinary blokes'; Ian is often mistaken for a school-teacher while Keith has something of the mien of a folk-singer. Both are family men: Ian's three children are now grown up, while Keith's two daughters are still at primary school. The cliché which says that successful scientists are ruthlessly ambitious also seems somewhat out of place. Ian stayed on at Roslin (then ABRO) in 1982 when the tide seemed to be going against him, largely because he and his wife, Vivienne, liked the area – and if he had not, the project that produced Dolly may not have come about. He says of his own ambition, 'I think I would like to be appreciated.' I have been in Keith's company while he and his partner Ange decided to turn down a lucrative offer to lecture in the United States – because it seemed to clash with the children's half-term holiday. Merciless ambition, one feels, should be made of sterner stuff.

This does not mean, however, that Ian and Keith do not take what they do seriously. Theirs, after all, will surely prove to be some of the most serious science and high technology to emerge from the 20th century, which is a considerable claim. Its implications will be felt through all the centuries to come.

We – all three of us – hope you enjoy this book. The story is complicated, but we believe it's worth the effort.

Colin Tudge

PART I

The New Age Begins

CHAPTER I

The Importance of Being Dolly

Ian Wilmut:

Dolly seems a very ordinary sheep – just an amiable Finn-Dorset ewe – yet as all the world has acknowledged, if not entirely for the right reasons, she might reasonably claim to be the most extraordinary creature ever to be born. Mammals are normally produced by the sexual route: an egg joins with a sperm, to form a new embryo. But in 1996 Keith Campbell and I, with our colleagues at Roslin Institute and PPL, cloned Dolly from a cell that had been taken from the mammary gland of an old ewe, and then grown in culture. The ewe, as it happened, was long since dead. We fused that cultured cell with an egg from yet another ewe to 'reconstruct' an embryo which we transferred into the womb of a surrogate mother, where it developed to become a lamb. This was the lamb we called Dolly: not quite the first mammal ever to be cloned, but certainly the first to be cloned from an adult body cell. Her birth overturns one of the deepest dogmas of all in biology for, until the moment in February 1997 that we made her existence known through a brief 'letter' in the scientific journal *Nature*, most scientists simply did not believe that cloning in such a way, and from such a cell, was possible. Even afterwards, some doubted that we had done what we claimed.

Dolly's impact was extraordinary. We expected a heavy response – the birth of Megan and Morag in 1995 had provided some warning of what might follow – but nothing could have prepared us for the (literally) thousands of telephone calls, the scores of interviews, the offers of tours and contracts, and in some cases the opprobrium – though much less of that than we might have feared. Everyone, worldwide, knew that Dolly was important. Even if they did not grasp her full significance (and the full significance, as this book will reveal, is not obvious and is far more profound than is generally appreciated), people the world over felt that life would never be quite the same again. And in this they are quite right.

Most obviously – and unfortunately, because it is certainly not the most important aspect – commentators worldwide immediately perceived that if a sheep can be cloned from a body cell then so can people. Many hated the idea, including President Clinton of the United States who called for a worldwide moratorium on all cloning research. But others welcomed human cloning and some – like Dr Richard Seed, who in fact is a physicist, not a physician – even offered to set up cloning clinics; though this is surely jumping the gun by several decades since very few scientists have the necessary expertise and, even in the best hands, human cloning at this stage would be absurdly risky. I fielded many of the telephone calls that flooded into Roslin Institute in the days after we went public with Dolly, and quickly came to dread the pleas from bereaved families, asking if we could clone their lost loved ones. I have two daughters and a son of my own and know that every parent's nightmare is to lose a child, and what you would give to have them back, but I had and have no power to help. I suppose this was my first, sharp intimation of the effect that Dolly could have on people's lives and perceptions. Such pleas are based on a misconception: that cloning of the kind that produced Dolly confers an instant, exact replication – a virtual resurrection. This simply is not the case. But the idea is pervasive, and was reflected in articles and cartoons around the world. The cover of *Der Spiegel* showed a regiment of Hitlers.

Yet human cloning is very far from Keith's and my own thoughts and ambitions, and we would rather that no one ever attempted it. If they do – and somebody, sometime surely will – it would be cruel not to wish good luck to everyone involved, but the prospect of human cloning causes us grave misgivings. It is physically too risky, it could have untoward effects on the psychology of the cloned child, and in the end we see no medical justification for it. For us, the technology that produced Dolly has far wider significance. As the decades and centuries pass, the science of cloning and the technologies that may flow from it will affect all aspects of human life – the things that people can do, the way we live, and even, if we choose, the kinds of people we are. Those future technologies will offer our successors a degree of control over life's processes that will come effectively to seem absolute. Until the birth of Dolly scientists were apt to declare that this or that procedure would be 'biologically impossible' – but now that expression seems to have lost all meaning. In the 21st century and beyond, human ambition will be bound only by the laws of physics, the rules of logic, and our descendants' own sense of right and wrong. Truly, Dolly has taken us into the age of biological control.

Dolly is not our only cloned sheep. Megan and Morag were our first outstanding successes – Welsh Mountain ewes cloned from cultured embryo cells. Taffy and Tweed, two Welsh Black rams, were cloned from cultured fetal cells at the same time as Dolly – and are at least as important as she is, since fetal cells may well be the best kind to work with. Were it not for Dolly, Taffy and Tweed would now be the most famous sheep in the world. At the same time as Dolly, too, we cloned Cedric, Cecil, Cyril and Tuppence from cultured embryo cells – four young Dorset rams who are genetically identical to each other and yet are very different in size and temperament, showing emphatically that an animal's genes do *not* 'determine' every detail of its physique and personality. This is one of several reasons why 'resurrection' of lost loved ones, human or otherwise, is not feasible.

But Keith and I did not set out simply to produce genetic replicas of existing animals. Some other biologists who have contributed enormously to the science and technology of cloning have indeed been motivated largely by the desire to replicate outstanding – 'elite' – livestock. But our broader and longer-term ambitions at Roslin, together with our collaborating biotech company PPL, lies in genetic engineering: the genetic 'transformation' of animals and of isolated animal and human tissues and cells, for a myriad of purposes in medicine, agriculture, conservation and pure science. Future possibilities will in principle be limited only by human imagination. A hint of what might come is provided not so much by Megan and Morag or by Dolly and her contemporaries, who have all been cloned but have not been genetically altered, but by Polly, born the year after Dolly, in 1997. Polly is both cloned *and* genetically transformed.

Indeed we should not see cloning as an isolated technology, single-mindedly directed at replication of livestock or of people. It is the third player in a trio of modern biotechnologies that have arisen since the early 1970s. Each of the three, taken alone, is striking; but taken together they take humanity into a new age – one as significant, as time will tell, as our forebears' transition into the age of steam, of radio, or of nuclear power.

The chief of these three biotechnologies is genetic engineering, which first began to be developed in the early 1970s. 'Genetic engineers' transfer genes from one organism to another – and, which is truly miraculous, the transferred genes may function perfectly in the new organism. The genetically engineered organism is then said to be 'transformed', or to be 'transgenic', and the transferred gene is called a 'transgene'. Some scientists and politicians in recent years have tried to underplay the significance of such gene transfer, suggesting that all it does is to accelerate the techniques of crop and livestock improvement that farmers and breeders have practised for thousands of years. Not so: traditional breeders must operate within the reproductive boundaries that define species. If they want to improve sheep, then they have to cross-breed the animals with other sheep. Potatoes can be improved only by crossing them with other potatoes. The modern genetic engineer, however, can in principle take genes from *any* organism and put them into any other: fungal genes into plants, mouse genes into bacteria, human genes into sheep. Again we see that traditional breeders were bound by the restraints of biology, whereas modern genetic engineers are in theory bound only by the laws of physics, by their imagination, and by the laws and ethics of their society. Genetic engineers have a precision, too, which traditional breeders lack: they can add just one gene at a time – or they take out individual genes, or take them out and alter them and put them back, or indeed (in principle) create quite new genes that have never existed before in nature. In evil hands such power could be ghoulish. Ethically directed, the potential for doing good is immense.

Genetic engineering, however, has been severely limited by the simple fact that most of the genes in most creatures remain unidentified. Thus human beings each have about 60 000–80 000 functional genes – but of these only a few thousand are known, what they look like, and what they do. So genetic engineers are developing the power to transfer genes from one organism to another but, for the most part, they do not know which genes to transfer. Over the past few decades, then, the science and technology of *genomics* has developed: the attempt to map all the genes in an organism, and eventually to unravel their individual structure and to find out what each of them does. The genes of some simple organisms – yeasts, the cress-like *Arabidopsis*, and the roundworm *Caenorhabditis* – have already been mapped in their entirety. Biologists worldwide are now cooperating to identify all the genes in the human being – this is the Human Genome Project, or HUGO. The first phase should be completed within another decade or so. Our colleagues at Roslin are now cooperating with other laboratories to identify and map all the genes in each of the common livestock species – poultry, sheep, cattle, pigs. When the knowledge gained by genomics comes on line, the power of genetic engineering will truly become evident.

Yet one player is missing. It is easy (relatively speaking!) to transform bacteria genetically. Put crudely (although in truth the procedures are immensely complicated) you just have to grow the bacteria in a dish and add DNA (DNA being the stuff of which genes are made) and then pick out the individual bacteria that have taken up the added genes most satisfactorily. The same, broadly speaking, can be done with plants. Plant tissue can be grown in a dish – which is what 'culturing' means; DNA is added (by various techniques that need not delay us here), and a whole new plant is regenerated from the cells that have taken up the added gene most effectively.

But with animals up until now – up until Polly, in fact – this just has not been possible. Genetic engineering of animals was first achieved in the 1980s and many animals have been genetically transformed since then – mostly laboratory mice but also more commercial species, such as cattle. But the only way to do this was to inject a gene (that is, a piece of DNA) into the young, one-celled embryo (otherwise known as a 'zygote') that is first formed by the fusion of egg and sperm. Then, with luck, all the cells of the animal that develops from that zygote will contain the new gene. This procedure has produced some remarkable results; notably, before Keith joined us at Roslin, I and various colleagues spent much of the 1980s putting human genes into sheep, so that they would produce valuable therapeutic proteins in their milk. As described in the next chapter, this became a serious commercial proposition with huge medical implications, as PPL is now demonstrating. Nevertheless, injection of DNA into zygotes is inefficient. How much better it would be to grow animal cells in a dish, as if they were bacteria or cultured plant cells; and then transform them *en masse;* and then – as is already carried out with bacteria and plants – grow whole new animals from the cells that had taken up the new genes most efficiently! Indeed with the cells already in culture genetic engineers are not confined simply to the addition of genes; they can subtract genes or alter them, or add artificial genes, just as is now possible in principle in bacteria and plants.

Until we started cloning sheep at Roslin, however, it simply was not possible to re-create whole animals from cultured cells. Keith came to Roslin in 1991 and he and I first achieved this in 1995 with Megan and Morag – who really should be seen as the most important of all our clones, for they were the ones who first showed that cloning from cultured cells is possible. Dolly, born in 1996, might be seen as the gilt on the gingerbread, although she had an added and stunning refinement – that she was grown from an adult cell. Polly, born in 1997, shows the promise of times to come. She was cloned from cultured cells which were transformed genetically – a human gene was added to them – as they were cultured.

But we will come to this. The point here is that the three technologies together – genetic engineering, genomics and our method of cloning from cultured cells – are a very powerful combination indeed. Genetic engineering is the conceptual leader: transfer of genes from organism to organism, and the creation of quite new genes, makes it possible in principle to build new organisms at will. Genomics provides the necessary data: knowledge of what genes to transfer – where to find them, and what they do. Cloning of the kind that we have developed at Roslin and PPL makes it possible in principle to apply all the immense power of genetic engineering and genomics to animals. Animals are the creatures that human beings identify with most closely: livestock form one of the most important components of the world's economy, and indeed of its ecology, and human beings, of course, are animals too. Commentators at large were right to observe that in principle, whatever can be done in sheep might also be done in people; but they did not for the most part perceive that cloning *per se* – mere replication – is only a fraction of what might in principle be done.

Yet these technologies, powerful as they may seem, are still not the end of the matter. Beyond technology, and in harness with it, is science. People conflate the two: most of what is reported on television by 'science' correspondents is in fact technology. Technology is about changing things: providing machines and medicines, or altering our surroundings, to make our own lives more comfortable and to create wealth. Science is about understanding how the universe works and all the creatures within it. The two pursuits are different, and not necessarily linked. Technology is as old as humankind: stone tools are technology. People may produce fine instruments and weapons, cathedrals, windmills and aqueducts, without having any formal knowledge of the underlying science – metallurgy, mechanics, aerodynamics and hydrodynamics. Contrariwise, science at its purest is nothing more nor less than 'natural philosophy', as it was originally known, and need produce no technologies at all.

Our method of cloning – transferring a nucleus from a body cell into an egg – is indeed a powerful technology but it also provides wonderful opportunities for scientific insight. Notably, biologists already have a good idea of how genes work but they would dearly like to know more. We know, for example, that genes make all creatures the way they are; they provide the proteins that form much of our body structure, and catalyse the reactions of the cell's metabolism. But we also know that the genes do not operate in isolation. They are in constant dialogue with the rest of the cell, which in turn responds to signals from the other cells of the body, which in the end are in touch with the world at large. The influence of factors outside the genes, which act upon them throughout life, is clearly seen in Cyril, Cedric, Cecil and Tuppence: four very different though genetically identical individuals.

The dialogue between the genes and their surroundings is understood to some extent, but we need to know far more. This dialogue controls the development of an organism from a single cell into a sheep – or indeed into a human being or an oak tree. It determines that some cells within an animal form brains, while others form liver or lung or a hundred other tissues; in other words, the dialogue shapes the processes of *differentiation*. Birth defects are sometimes caused by flaws in the genes themselves – harmful mutations – but they also result from interruptions in the dialogue between the genes and their surroundings, for example by toxins or infections. The dialogue between genes and their surroundings continues after the animal is born and throughout life, and if it goes awry the genes go out of control; the cells grow wildly and the result is cancer. In short, once we understand how the genes interact with their surroundings – the nature of the dialogue – then, truly, we will begin to appreciate how bodies really work, and develop, and what goes wrong in disease. That understanding is science.

But although science and technology are different pursuits with different histories, they work in concert. Technology without science is, well, technology: stone tools, windmills, mud huts. Technology with science is 'high technology'; 'high-tech' is the technology that emerges from science. 'Biotechnology' is high-tech of a biological nature: genetic engineering and cloning are the prime examples. Biotechnology is rapidly becoming one of the world's great industries. In truth, however, ideas do not flow simply from science into technology. The flow runs in both directions – for without technology, science itself would grind to a halt. We shall see throughout this book how the science and craft of cloning depends on technological input: extraordinary microscopes of wonderful optical purity, high-precision instruments for microdissection, preparations of purified hormones to control the reproductive cycles of our experimental animals, methods for genetic analysis, and so on.

Both Keith and I, in our pursuit of cloning, are engrossed both in the science and in the technology that makes cloning possible and which will develop from it in the future. I once wanted to be a farmer and am very happy with the idea that scientific research should have practical results – that it should indeed lead to useful high technologies. My current interest in medical biotechnology was fired by the suffering of my own father, who was diabetic, was blinded by the disease in the 1960s, and lost part of a leg and much of the use of his hands before his death in 1994. As we will see later, we intend to adapt aspects of the technology that produced Dolly to provide a cure for diabetes: it will be possible one day to replicate and restore the islet cells that produce insulin, the hormone that diabetics are lacking. The same principle can be applied to many other diseases. Keith perhaps is more of a pure scientist – 'I just want to know how everything works,' he says. His curiosity started young: as a boy, he filled his mother's kitchen with frogs. But he has also been a medical technician, and has worked on cancer; and after Dolly was born he moved from Roslin, the scientific research laboratory, to PPL, the commercial biotech company, and now on to Nottingham (my old university) to be a professor. In reality the scientific research and the biotechnological development run into each other.

All this then, in a nutshell, is the true significance of our work. This is why Dolly matters – and why Megan and Morag, Taffy and Tweed, and Polly, perhaps matter even more. Human cloning has grabbed people's imagination, but that is merely a diversion – and one we personally regret, and find distasteful. We did not make Dolly for that. Still less did we ever intend to produce vast flocks of identical sheep. If all you want to do is multiply sheep, then good old-fashioned sex is the way to do it. Our work completes the biotechnological trio: genetic engineering, genomics and cloning. It also provides an extraordinarily powerful scientific model for studying the interactions of genes and their surroundings – interactions which between them account for so much of development and disease. Taken together, the new biotechnologies and the pending scientific insights will be immensely powerful. Truly they will take humanity into the age of biological control.

So how did we come to do all this? If we simply wanted a good story, we could tell it as if Dolly was our destiny. The research that produced Megan and Morag, and then Dolly and the rest, requires two main kinds of expertise: embryology or developmental biology on the one hand – the science that describes how sperm and eggs normally combine to make single-cell embryos, which multiply and grow to form entire organisms, like us and sheep, with billions of cells of many different types – and cell biology on the other – a knowledge of the workings of individual cells, how they grow and divide, and how they can be manipulated. I set out from the start of my scientific career to be an embryologist and Keith was always a cell biologist, and both of us were interested in cloning at least from the 1980s. I first came to Roslin as a senior scientific officer in 1973 (though it was called ABRO in those days) and began the programme that eventually led to Dolly after 1986; Keith joined us with just the expertise we required, and some revolutionary ideas, in 1991. I suppose there is a hint of destiny about all this. In truth, though, the outcome of science is not predestined. Luck plays a large part, and so does serendipity, which is something more than luck: unexpected, and unlooked-for happiness. To a large extent, luck and serendipity made us the kind of scientists we are, and brought us together. The cliché has it that if we had not done this work then somebody else would soon have done so. But this common presumption comes with no guarantees. You can judge for yourself, as the story of Dolly unfolds, whether others would have gone down the same path – and if so, how quickly. History can only tell us what happened; it cannot tell us what might have been.

We are very different people, Keith and I. We have got along well this past decade, but although we have had many an amicable and intense conversation in our offices at Roslin we very rarely meet outside of work even though we live within a few miles of each other. We both have families to get home to. I am now in my fifties, and Keith in his forties. We have both worked in Scotland for some time – I since the early 1970s, while Keith had research fellowships at Edinburgh and Dundee Universities before he joined Roslin (then called IAPGR) in 1991 – and so we almost qualify as honorary Scots; but we both came originally from the Midlands of England. My father was a maths teacher, a man of immense presence (though not physically large) and I greatly admired him. As already noted, he became diabetic at an early age and later – in the 1960s – the disease made him blind. He bore his suffering patiently, and became a computer programmer after the blindness made it impossible to teach in school. Perhaps I resemble him a bit in character; I would like to think so. People often seem to think I'm a schoolteacher. Perhaps it's the beard and the bald head, and the fact that I do like to be in charge. Or, at least, I don't like other people telling me what to do. I suppose it goes with the job.

Keith and I do different things. I live with my wife Vivienne in a village some way to the south of Roslin – our children are well in their twenties and have flown the nest – while Keith lives even further away with his partner, Ange, and their two small daughters. So we commute. Science can be an extraordinary job with extraordinary outcomes, but for the most part daily life is as routine as any other. We have to clock in these days. We spend much of our time applying for grants. For relaxation I used to jog. Now I am content to walk over the hills. I like to belong to the local community and have become reasonably adept at curling, the peculiarly Scottish game (though it is now an Olympic sport much favoured by the Japanese) which resembles bowls on ice. The 'bowls' in fact are stones, big and flat like old-fashioned kettles, though a lot heavier, sent skittering over the ice. I don't like to compete, though. I get too nervous. Keith has a more arty look. He plays the drums, and races at high speed over the hills on a mountain bike. His daughters are still young, too, which takes up his time.

I did not exactly shine at school – though I did meet Vivienne at that time. Her grammar school was next to mine, although the authorities had thoughtfully positioned the gates as far apart as possible to reduce contact to a minimum and it was a long walk from one to the other. I seemed to work reasonably hard but left school with too few passes at A level to go to university and had to do a catch-up year before I joined the agricultural college at Nottingham University. Farming was my first love – but to be a successful farmer you have to be good at business; and I soon realised that I am not.

So I shuffled sideways into scientific research. In my last year as an undergraduate, in the long summer vacation of 1966, I won a scholarship from the Pig Industry Development Authority (PIDA) to work for eight weeks under E. J. C. (Chris) Polge at the ARC's Unit of Reproductive Physiology and Biochemistry at Cambridge (usually just called 'the Animal Research Station'). Chris is a wonderful man: an excellent scientist (he is a Fellow of the Royal Society) but also kind and jovial. My job was to assist generally with the experiments, and there I learned to work with embryos. I found them very beautiful, and so those weeks with Chris set the cast of my life. Chris later became one of the pioneer scientist–entrepreneurs of the 1980s, when he established a company called Animal Biotechnology Cambridge.

When I graduated from Nottingham in 1967 (with an upper second: not bad, not brilliant) I went back to work with Chris at the Animal Research Station. He supervised me through my PhD – on the freezing of boar semen – which was awarded by Darwin College, Cambridge, in 1971. I stayed on with Chris, continuing the freezing research, after I got my doctorate; and so in 1973 I became the first scientist to freeze a calf embryo successfully, thaw it again, and transfer it to a surrogate mother – who went on to give birth to the world's first 'frozen calf', a red-and-white Hereford–Friesian cross whom I called 'Frostie'. The *Veterinary*

Record published the news in June 1973 within weeks of his birth, which was one of the fastest scientific publications ever. Frostie gave me my first taste of the media. Newspapers as far away as New Zealand picked up the story, while Britain's *Daily Mail* won the perennial headline competition with 'Ice-Age calf weighs in'. I did my first TV interview, with the BBC at Norwich, relayed to London. Frozen calf embryos have since played a significant part in agriculture and even in conservation. Frozen embryos of kudu antelope have been transferred into the wombs of the more common eland, and born successfully.

In short, with Chris Polge in the 1960s and 70s, I learned many of the basic techniques of reproductive physiology that later fed in to our cloning work at Roslin. As we will see, another key figure in the history of cloning is the Danish veterinarian Steen Willadsen, who among other things gave me some vital tips in the mid 1980s. He too worked for Chris Polge – in fact he took over my job when I left. Willadsen began cloning at the Animal Research Station, initially just by dividing embryos. But more of this later. I left Chris's laboratory in 1973 – to join ABRO, which became IAPGR, which became Roslin Institute.

Keith is the son of a seedsman, and was brought up in part on the farm where his grandfather worked. He did not exactly shine at his grammar school either. He began work as a medical technician but, he says, 'I resigned my job the day I qualified to do it' – and went on to read microbiology at Queen Elizabeth College, London. He says he was lazy at university but every undergraduate claims as much, and in fact he attended all the lectures and wrote up the notes conscientiously. He too got an upper second.

Keith, by training and inclination, is a cell biologist. By the standards of most scientists his career (as is often commented) is 'unusual'. After he graduated from Queen Elizabeth he worked for a time in Yemen as a medical technician; then in Sussex helping to control Dutch Elm disease; and then for the Marie Curie Institute in Surrey. There he saw how cancer cells change their character – from specialist, differentiated cells into less differentiated cells, and they then may differentiate again in new ways; an insight which, later, deeply affected his approach to cloning. He completed his doctorate (a DPhil, the same as a PhD) at Sussex University in 1987 with a thesis 'Aspects of cell cycle regulation in yeast and *Xenopus*' – *Xenopus* being the African clawed frog, which featured in some of the very earliest cloning experiments in the 1960s. Again, this experience profoundly influenced his approach to cloning because, he says, 'I realised that when you look closely at cells, they are all much the same. If yeasts can clone – which they do – I didn't see any absolute reason why mammals couldn't be cloned too.' Note, too, Keith's interest in 'the cell cycle'. We will discuss what this means in chapter 8; it seems to be the key to cloning success. As a 'postdoc' Keith first spent two years at the University of Edinburgh, then another 18 months at the University of Dundee before seeing the advertisement that I had placed in *Nature*, in February 1991, which lured him to Roslin. By then, I was already several years into the cloning project, and knew that I needed a cell biologist who understood cell cycles. Keith seemed tailor-made for the task.

Science is a human pursuit. Style matters. Keith and I have different backgrounds as well as styles, both of which I believe are complementary; it has been a fruitful alliance. Thus it is vital, in science, to be sure of your ground; if you once start assuming things that are not clearly established then you may follow a false trail that may take the rest of your career. There are many examples of this. You cannot be certain of everything in science, but you have to be as certain as you can of the things that are knowable. On the one hand, my own philosophy, and *modus operandi,* has been to identify the broad nature of the problem in hand, for as long ahead as possible, then to identify step by step the things we need to know in order to provide a solution, and then to move methodically through the steps, leaving as little as possible to chance. Such was the route I planned from the late 1980s to take us into cloning. Keith on the other hand is more adventurous. It was only when Keith joined the group and brought in his different experience that I fully understood the significance of the cell cycle in the development of cloned embryos. If our different experiences had not blended so well we certainly would not have produced Megan and Morag, and then Dolly, by the middle of the 1990s. The science needed a change of direction, a new focus.

In this book Keith and I want to explain the full story of how and why we came to clone first Megan and Morag and then Dolly, who led on to Polly. As scientists and as citizens we do not want to be cut off from the rest of society. We do not want to be doing strange and arcane things entirely out of the public gaze. In a democracy, science and technology need the overall approval of society; and people at large need to know what is happening, if society is to stay in control. Besides, we are actually rather proud of the work, and would like others to appreciate it.

The full story is, however, inescapably complicated. The science and technology of cloning, at least by our method, takes us into some of the most esoteric reaches of biology: the details of reproductive physiology, the intricacies of gene behaviour – and overall, as the great Russian–American biologist Theodosius Dobzhansky once commented, 'Nothing makes sense in biology except in the light of evolution', so we should look at the broad, philosophical background as well. We have made the explanations as simple as we can without fudging and garbling, but as Einstein said, 'Explanations must be as simple as possible – *and no simpler.*' We have not ducked the technical language. Technical terms describe entities and phenomena that simply do not exist in everyday life, and unless we employ those terms we cannot refer, clearly, to the essential elements of the story. So we will explain what the terms mean, and then use them. 'Karyoplast', 'cytoplast', 'maturation-promoting factor' (MPF), and the various components of the cell and the cell cycle are a few (actually the most important) of the terms that appear throughout this book. We also believe that one of the main points of 'popular' accounts such as this is to introduce non-specialists to the real thing – so that, with luck, when you have finished reading this book you will be able to read the original, specialist accounts, with understanding and even with pleasure. To see whether we have succeeded in this we are presenting one of the key papers – the one that described Dolly – at the end of the book.

But although the story is complicated it is biology, and biology is not physics: it is not *weird* – it does not ask you to believe that time passes at different rates in different circumstances, or that a photon can be in two places at once – and it is not so rooted in maths, which in general the human brain does not do well. So although biology can be complicated there is nothing in it that anyone who picks up this book cannot understand; and we think the research is worth understanding both for the cultural reason – that science is interesting – and for the practical reason – that the technology that emerges from the science affects all of us, in a hundred different ways. A complicated narrative, however, needs a preview, to give shape to it. So in the rest of this introductory chapter I will give a rough outline of all that we did, and why.

When I came to Roslin (ABRO) in 1973 it was not to do cloning, or anything like it. As we will see in later chapters, Robert Briggs and Thomas King in America and John Gurdon in Cambridge had carried out the first cloning studies in frogs by that time, but cloning of mammals was not properly on the agenda until the 1980s. Even if it had been, I had no reason to be involved in it at the start of my career. At first I worked on a variety of developmental problems and by the start of the 1980s I was carrying out research into embryo death. In farm animals about twenty-five per cent of eggs that are fertilised fail to produce live births. Yet other animals do better: mice, for example, generally lose twenty per cent of their pups *in utero* but strains can be bred that lose only about ten per cent; and animals such as red deer that have only a very short breeding season, and in effect cannot *afford* to lose embryos, have adapted accordingly and lose far fewer than cattle do. It was, and is, a fascinating problem and an important one.

But in 1982, just as I was getting seriously to grips with embryo death, I was obliged to change course. The Agricultural Research Council, which was then the government agency in charge of ABRO, decided that government-supported research in general needed to change its philosophy, and that most of the research stations had to trim their sails. The fashionable science at that time was molecular biology; and John King, who was then director, wanted to introduce more molecular biology – which required more funds, rather than less. King was shortly succeeded as director by Roger Land, who brought in John Bishop from the department of genetics at Edinburgh as a special consultant. ABRO then appointed its first full-time molecular biologist, Rick Lathe, who worked at the biotechnology laboratory Transgene in Strasbourg, France.

Bishop and Lathe then came up with the idea that transformed much of the research at Roslin, gave rise to PPL, and brought me into the line of research that has led to Dolly. Again, however, on the face of things, their idea seemed to have nothing directly to do with cloning. They wanted to carry out genetic engineering in sheep. Specifically, they wanted to see whether it would be possible to transfer human genes into sheep that would make proteins of therapeutic value. Among many they had in mind were proteins involved in blood clotting, notably factor VIII and factor IX, a lack of which leads to haemophilia; and the enzyme alpha-1-antitrypsin, or AAT, used to treat sufferers from emphysema and cystic fibrosis.

Again, this work seemed to have nothing much to do with me or my line of interest, but Roger Land decided to bring my work on embryo death to a close and offered me the chance to work on the genetic engineering project instead. As we have seen, genetic engineering in those days (and for the most part, even in these days) was carried out by injecting DNA into one-celled embryos (zygotes); my job would be to supply the zygotes, and indeed to inject the DNA provided by the molecular biologists. Supply of zygotes certainly called upon my skills as a developmental biologist but injection of DNA did not. For one thing, I have a problem with very small scale manipulations because I have a hand tremor. For another, this work seemed tediously routine. It did not present the kind of intellectual challenges that had attracted me into science. So at that time I thought long and hard about leaving ABRO. But 1982 was a hard year for British agricultural science. Many of my colleagues had simply lost their jobs. I discussed the whole issue at length with Vivienne and decided I should tough it out. We had been in Scotland for nearly 10 years, we liked the area and the people, the children were at school, and life is rarely perfect; sometimes you just have to put up with it. From the start, though, I agreed with Roger Land that as soon as I had developed a routine for genetic transformation of zygotes I should seek more efficient methods.

This early work on genetic transformation of sheep at Roslin – without cloning – is described in chapter 2. Out of that work came Tracy, born in 1990, who expressed large amounts of AAT in her milk. PPL took over this work and now has flocks of sheep similar to Tracy that produce commercial quantities of AAT. This is currently undergoing clinical assessment in Britain, and if all goes well will be commercially available within a few years.

So where does cloning come into this? As we will see in the next chapter, the standard way of making genetic transformations – by injecting DNA into the zygote – is unsatisfactory in several ways. One obvious drawback was (and is) that there is only one zygote per animal, so you get only one attempt per animal; if that fails, you waste a whole zygote. It soon occurred to me that it would be better if we could first allow the zygote to multiply, to produce several or many cells; and then add new DNA to several or many of those cells; and then produce new embryos from each of the transformed cells. Such multiplication *is* cloning.

Research elsewhere in the early 1980s suggested a possible route. Biologists working on mice had lines of embryo stems that retained embryonic qualities when they were multiplied in culture. These were called embryo stem cells, or ES cells. ES cells could be taken out of culture and then put back into more mouse embryos – and then they were able to develop into any of the tissues of the host embryo: muscle, brain, gut, or indeed eggs or sperm. Thus mouse biologists were able to carry out genetic engineering with ES cells: they could transform the cells in culture, then return the transformed cells into mouse embryos, and then hope that some of the transformed cells would form eggs or sperm so that when the embryos grew into adult mice they would produce genetically transformed offspring.

Note, however, that the mouse biologists of the early 1980s were never able to create whole new mice directly from ES cells, transformed or otherwise – and, indeed, this has never been done. But they were able to add transformed cells to existing embryos. I felt that if only we could create ES cells from sheep then we could do the same thing. But a few years later – after a meeting with Steen Willadsen in 1986 – I realised that in sheep we should be able to go one better than the mouse biologists; we should, in fact, be able to create embryos directly from transformed ES cells. In other words, we should be able to clone sheep from ES cells – and if we transformed the ES cells first we would thereby produce genetically transformed sheep.

So I first moved into genetic engineering of sheep and then sought to apply the technology of cloning to this endeavour – by creating, and then cloning, ES cells. This ambition set the tone of my research from the late 1980s to the early 1990s. Its complexities will unfold in the following chapters. But, as we will also see, the method that eventually led to success, with Megan and Morag and then with Dolly, did *not* involve ES cells; there is a huge twist in the story. Science is a logical pursuit but progress in science does not necessarily, or even usually, follow a straight path. You just have to follow your nose – and sometimes you must put your treasured ideas aside, and go on down a different route. The different route, in this case, was opened up by Keith.

But I am running ahead. We should take the story step by step. It is appropriate to begin with the line of work that first got me into genetic engineering and then led, by a circuitous route, into cloning. We should begin, in short, with Roslin's (then IAPGR's) first great success in this field: Tracy.

CHAPTER 2

Tracy: The Most Valuable Milk
in the World

Ian Wilmut:

Tracy is not cloned, but she is genetically engineered: one of the first transgenic animals of truly commercial significance. I was involved in the research that produced her, but was well into other things – cloning – by the time she was born in 1990. She is highly pertinent to our story, however. The research that produced her inspired the work that led to Megan and Morag and to Dolly; cloning, or at least our version of it, is primarily conceived as an aid to genetic transformation. Polly, born the year after Dolly in 1997, combines the kind of cloning techniques that produced Taffy and Tweed with the methods that produced Tracy. So we must talk about Tracy.

Tracy has been fitted with a human gene that produces the enzyme alpha-1-antitrypsin (commonly abbreviated to AAT) and she obligingly secretes enormous quantities in her milk. AAT is already used in the United States to treat lung disorders, notably emphysema and cystic fibrosis, but present supplies come from human blood plasma and so are both limited and expensive, while any human blood product carries a theoretical threat of infection. As much AAT as is needed could in principle be produced from sheep's milk more safely and much more cheaply – and many another therapeutic protein besides. This method of production has been called 'pharming' – a felicitous pun. Tracy herself is now

35

an old lady and somewhat creaky but sheep like her, and their many descendants, continue to secrete the enzyme as they lactate. Milk-derived AAT is now undergoing clinical trials in Britain to treat cystic fibrosis (it is delivered to the lungs by aerosol) and PPL should be marketing it in 2001.

The ideas and vocabulary that relate to genes in general and genetic engineering in particular surface throughout this book so we should take this opportunity to bring them into the discussion.

Genes and genetic engineering: a brief primer

The twin seeds of genetic engineering were sown within a decade of each other in the mid 19th century. First, in the 1860s, Gregor Mendel working in what is now the Czech Republic showed that the individual features – 'characters' – of plants were 'determined' (slightly too strong a word, but it will do) by discrete 'factors' which later, in the 20th century, were given the name of 'genes'. Of course – as Mendel knew full well! – *most* characters are not determined by single genes. Most are *polygenic*, which means they are brought about by combinations of many different genes; and – just to stir the pot a little more – most genes are *pleiotropic*, which means that they affect more than one character, often characters that seem quite unrelated to each other. The fact that most characters are polygenic and most genes are pleiotropic severely limits the scope and power of genetic engineering, at least in the present state of knowledge. 'Genetic engineers' can transfer only one gene at a time and want its effects to be entirely predictable – which means that in general they must focus upon the *minority* of characters that have a simple genetic basis, and are determined by genes that are not highly pleiotropic.

A few years after Mendel, in 1869, and quite separately, a Swiss biochemist called Johann Friedrich Miescher working at Tubingen in Germany discovered DNA in pus cells. He called it 'nuclein' (the name 'nucleic acid' was coined much later, in 1889, by another scientist called Richard Altmann) and in the 1870s he studied nuclein in the sperm of Rhine salmon and concluded that it might be 'the specific cause of fertilisation'. But Miescher died in 1895, aged just 51, and never really followed up what in truth was a fine insight.

So Mendel showed the underlying mechanism of heredity and gave rise to the science of 'classical genetics' (although he did not coin the terms 'genetics' or 'gene'); and Miescher discovered the material of which genes are made (although he did not realise this at the time); and so classical genetics (which treats genes as abstractions) was able to become 'molecular genetics' (which treats genes as pieces of DNA). In fact, of course, the history of science is human history and human history never runs as logically as this. For DNA seemed to be chemically boring stuff, with very little variation; but genes are obviously extremely various. So most biologists in the early 20th century felt that genes could *not* be made of DNA. Genes must surely be made of something that is itself various – and the most obvious candidate was protein. In the end it was not until the 1940s that Oswald Avery and his colleagues in the United States showed beyond all doubt that DNA is indeed the stuff of which genes are made. Then biologists finally perceived that DNA, superficially dull though it may seem to be, is in fact as James Watson has said 'the most interesting molecule in all of nature'. By the 1940s, too, the Scottish biochemist Alexander Todd and others had worked out the basic chemistry of DNA – what is in it – and in the early 1950s James Watson and Francis Crick showed how all the various bits fit together. In 1953 they produced their wonderful three-dimensional model: the famous 'double helix'.

Even before genes were known to be made of DNA – in fact from the beginning of the 20th century – it was clear that the principal task of genes is to make proteins (so it is hard to see how they could be proteins themselves). In particular, many proteins function as enzymes, and the task of the enzymes is to act as catalysts and so to control the cell's metabolism, and to help to synthesise all the many other molecules of which cells are composed (such as complex fats). Hence the genes, operating *via* proteins, control both the structure and the function of the cells. Of course they are not in absolute control; in reality genes act in dialogue with their surroundings – a theme that recurs again and again in this book and is crucial to a true understanding of cloning. But the genes certainly set the general direction of events, and determine the limits of what any one organism can do.

The conceptual problem is – or was – that any one body produces many thousands of different proteins, which do hundreds of thousands of different things, but DNA itself seems chemically simple. In fact a DNA molecule has only three basic components: a sugar called deoxyribose; a number of phosphorus-containing groups called 'phosphate radicals'; and a set of four 'bases' or 'nucleotides' known as adenine (A), cytosine (C), thymine (T), and guanine (G). These four bases provide the *only* source of variation in the DNA molecule. No wonder biologists thought it was boring, and could not possibly be the stuff of genes! How could such simplicity generate such complexity, and with such precision? But, as always, nature is way ahead of us. As became evident in the early 1960s, the *order* in which the four bases occur in the DNA molecule provides a 'code' that is in principle rich enough to specify all the proteins that any living thing could ever require: an infinity of different possibilities.

For proteins, like DNA, are made up from chains of subunits; in the case of proteins, these are chains of amino acids. Only about twenty amino acids occur in nature (or, at least, in animals) but they can occur in any order along the chain. The chains can also be as long as you like, and after the chains are completed they are folded carefully into intricate three-dimensional structures, and typically they are then adorned with sugars and metal ions to refine their function. So, although there are only twenty or so amino acids, the range of proteins to which they can give rise is effectively infinite. In the same way the twenty-six letters of the western alphabet can code the language of Shakespeare, and 10 000 other languages as well.

Francis Crick and Sydney Brenner thought that the task of 'cracking' the genetic code would take generations but in truth they hit upon the basic principle almost immediately. Logically, they said, each different amino acid should be coded by its own particular sequence of bases along the DNA molecule. How long, or short, would these sequences logically need to be to do the job? If each amino acid was coded by a sequence of only two bases, then there would not be enough variation: since there are only four different bases there are only sixteen possible two-base sequences (four times four). What if each amino acid was coded by a three-base sequence? Then there are sixty-four possible variants – or four times four times four. They tried this, and found that lo!, what was logically the simplest solution is in fact what nature has chosen to do. Each amino acid *is* coded by a corresponding sequence of three bases on the DNA. They called each set of three a *codon*. The order of the codons on the DNA determines the order of amino acids in the protein that is made from it, which in turn determines what kind of protein it is. In fact, since there are only twenty amino acids and there are sixty-four possible different codons, the code has information to spare. As it turns out, some amino acids are specified by more than one codon (though no one codon ever specifies more than one amino acid!) while other codons do not serve as codes for amino acids but rather as punctuation marks, marking where each gene begins and ends. After all, the DNA molecule is one continuous chain of bases, and without punctuation, the code would be nonsensical. Thus, the order of play of four bases in a long molecule does indeed provide an organism with all the information it needs to do all the things an organism does. Astonishing!

There is one significant refinement – or anomaly – however, just in case we should be tempted to suppose that nature is too simple. Not all of the bases in a chain of DNA code for proteins, or act as punctuation marks. It is not the case that every piece of DNA functions as a gene. In fact, only about five per cent of each DNA molecule actually provides any meaningful code. The other ninety-five per cent *looks* like a fairly nonsensical sequence. Indeed, this ninety-five per cent has sometimes been written off as 'junk' DNA. This is, of course, a very peremptory judgement; it is highly presumptuous to suggest that what is not understood is therefore junk. It seems likely, however, that much of the ninety-five per cent *is* junk: old genes that have been half broken down and forgotten about; stretches of DNA that are good at getting themselves replicated but do not actually get round to making proteins; perhaps old bits of viruses that have got themselves integrated into the DNA molecule (for viruses are really just mobile bits of DNA, wrapped in a protective protein coat). Overall, however, it is clear that this alleged 'junk' affects the *expression* of genes – the extent to which they are actually allowed to operate. Intriguingly, some of the non-coding DNA (the term 'junk' is best abandoned) occurs *within* the coding sequences that act as genes, and these internal interruptions are termed '*introns*'. The coding stretches of DNA, somewhat perversely, are correspondingly called '*exons*'. Often the introns are larger than the exons. But much of the non-coding DNA occurs between the genes.

In truth, a DNA molecule should not simply be called a 'molecule' since it consists of many different molecules – bases, sugars, and phosphate radicals – all joined together. It should be called a 'macromolecule'. In practice, eukaryotic creatures like animals, plants and fungi keep most of their DNA snug within a nucleus. Each nucleus contains several or many DNA macromolecules. In fact each macromolecule forms a *chromosome*. A chromosome does not consist solely of DNA, however, for the DNA (which is already coiled to form a double helix) is supercoiled around a core which consists primarily of proteins known as *histones*, and provides a kind of skeleton for the chromosomes. Each species of animal or plant has a characteristic number of chromosomes: fruit flies have four, for example, humans have forty-six, chimpanzees have forty-eight, and sheep have fifty-four. Overall, there is no clear relationship between the number of chromosomes and body size, or IQ, or any other obvious parameter: chromosome number merely reflects the way a particular species happens to package its DNA. In addition, each animal or plant inherits half of a complete set of chromosomes from each of its parents – so that, when the new embryo is formed, it contains two complete sets. This is why each species always has an even number of chromosomes: human beings inherit twenty-three chromosomes from each parent, sheep inherit twenty-seven from each parent, and so on.

Finally, in this lightning tour of terms, note that the total complement of genes within any one creature is called its *genome*. Since the genes are carried on the chromosomes, and each body cell contains two sets of chromosomes, each body cell carries two complete copies of the genome. The two copies are liable to be somewhat different from each other, however, since although we inherit the same broad *kinds* of genes from each parent (the kinds that are characteristic of our species) each gene may come in one of several different versions, with each version being known as an *allele*. Thus we will inherit genes for human eye colour from each parent, but we may inherit a blue-coding allele from one and a brown-coding allele from the other.

One last point: on how genes actually function. DNA does indeed provide the code for proteins but it does not make proteins directly. Instead it sits in the nucleus and, with the aid of enzymes, produces a second kind of nucleic acid known as RNA (short for *ribonucleic acid*). This process is called 'transcription'. RNA consists of a chain of bases, as in DNA; but one of the bases is different (RNA contains uracil instead of thymine) and the sugars attached to the bases are also different (ribose instead of deoxyribose). Even so, the sequence of bases in each piece of RNA corresponds to the sequence of bases in the piece of DNA – the gene – that made it, introns and all, at this stage. Then the RNA leaves the nucleus and the copies of the introns are trimmed out of it. It travels to a structure within the cytoplasm of the cell known as a 'ribosome'. The ribosome contains yet more RNA of a different kind (also made by the DNA, of course) and its job is to act as a kind of universal protein-making factory – a small factory, more like a blacksmith at his forge. The ribosome then makes a protein according to the code provided by the DNA that has travelled hotfoot from the nucleus. In short, as Francis Crick put the matter in what he called 'the central dogma' of molecular biology, 'DNA makes RNA makes protein!' DNA, RNA, and protein are indeed the trinity on which all life on Earth is based.

We could continue with this: the intricacies and refinements are truly wonderful. But we have said enough for present purposes. The key point is that Gregor Mendel established the existence of genes (although he simply called them 'factors') and out of this insight arose the 20th-century study of genetics. For the first half of the 20th century – which was the great age of 'classical genetics' – geneticists were obliged to treat genes purely as abstractions, and commonly envisaged them like beads on a string (where the chromosomes are the strings). But then, in 1953, Crick and Watson revealed the three-dimensional structure of DNA and so molecular biology was born; and now we understand in pleasing detail how DNA operates and what it does. Classical genetics continues – it is the basis of livestock breeding, genetic counselling, and studies in conservation – but now it is abetted by molecular biology. Geneticists can think of genes as beads on a string when it is helpful to do so – but they can also, if they choose, think of genes as real entities: lengths of DNA.

In short, with the birth of molecular biology, genetics could become an exercise in chemistry: highly refined chemistry, but chemistry nevertheless. The physical stuff of which genes are made, DNA, can be physically manipulated. Pieces of DNA can be bodily transferred from one creature to another, modifying the inheritance of the receiving organism as a mechanic might modify an engine. In short, out of the science of molecular biology has grown the technology of 'genetic engineering'.

Genetic engineering dates from the early 1970s; biologists found that pieces of DNA from different organisms could be joined together to produce 'recombinant DNA'. They achieved this with the aid of enzymes whose normal job is to edit and repair the DNA molecules within the cell nucleus. Then came a huge serendipity – perhaps the greatest in the whole history of biotechnology. In 1973 biologists found that they could introduce new genes into bacteria – in fact, the ubiquitous gut bacterium *Escherichia coli* (or *E. coli* for short) – simply by sprinkling them with DNA as they grew in culture. Extraordinarily, some of the bacteria took in the added DNA and (sometimes) incorporated it into their own genomes; and sometimes the newly integrated DNA was a fully functional gene. Organisms that have been altered genetically by such means are said to be *transformed* and after transformation they are *transgenic*. The gene that is transferred is a *transgene* and the act of transforming is *transgenesis*. 'Genetic engineering' is the popular term.

Before the age of genetic engineering genes could pass between organisms only through the formal mechanisms of sex – or occasionally, and effectively randomly, through the agency of viruses. In creatures other than bacteria, sexual exchange is in general possible only between creatures of the same or closely related species. But with genetic engineering it is in principle possible to transfer genes between *any* two organisms no matter how distant: from human being to bacterium, from plant to human being, whatever. Unsurprisingly, and very properly, genetic engineering in its early days gave rise to debates at least as cogent and emotionally fraught as those that have surrounded Dolly. President Clinton called for a moratorium: the Nobel laureate Paul Berg, himself a pioneer of genetic engineering, called for one in the early 1970s.

Plants are in some ways more difficult to transform than animals: they have thick cell walls, after all, and they cannot in general be transformed simply by sprinkling them with DNA. But they are in general more tolerant – for example of variations in chromosome number – and they do not raise problems of welfare, so the first successes were recorded in plants before the 1970s were out.

Animals were first transformed in 1980 by Frank Ruddle and Jon Gordon at Yale. By 1986 there had appeared a manual, *Manipulation of the Mouse Embryo*. Now, hundreds of animals of many species have been genetically transformed – mice, pigs, cows, sheep – by injecting their embryos with many different kinds of DNA. So what is entailed?

Animals transformed

Ruddle and Gordon in 1980 *injected* DNA into the nuclei of one-cell mouse embryos; and this has been the standard method of transgenesis in animals ever since – but also with a few important refinements, as we shall see. In practice (as we will see in chapter 5) a one-cell embryo (a zygote) does not have a single nucleus. Until the time at which it divides it has two *pronuclei* – one derived from the egg, and one derived from the incoming sperm. Animal genetic engineers inject the DNA into one or other of the pronuclei. If the DNA is incorporated immediately into one or other of the chromosomes within that pronucleus then copies of it will be passed on into all the cells of the body, since all the body cells derive from the initial zygote. In fact, a gene introduced into a zygote pronucleus should, if all goes well, be incorporated within the eggs or sperm of the new animal as it develops, and so will be passed on to all subsequent generations. This is *germ-line* transgenesis. It makes a permanent, heritable change.

Animal cells can also be transformed as they grow in culture – and in these circumstances it is sufficient simply to sprinkle DNA over the top, though there are extra ways of helping the DNA to enter (for example by incorporating it into tiny capsules of fat, which move more easily through the fatty outer membrane of the cell). Cells thus transformed might be reincorporated with the animal from which they came – and then might, for example, be used to repair tissue that has been damaged by some genetic defect. Such therapy has at least been mooted in cystic fibrosis, for example. But when transformed tissue is added to the body in this way the genes do not enter the eggs or sperm and are not passed on to subsequent generations. This is *ad hoc* repair, confined to the particular patient.

For 'pure' biologists transgenesis is a superb means of investigation: one very instructive way to find out what a gene can really do, for example, is to take it out of one genome and put it into another, and see what difference that makes. Animals can also serve as 'models' for human genetic diseases. We are considering making a strain of transgenic sheep containing the mutant gene that causes cystic fibrosis in humans. Sheep lungs are similar to those of humans, and they should provide a good model for the disease. At the moment, however, they lack the mutation that causes it.

But in addition, and very obviously, genetic engineering might simply be employed to improve the performance of livestock. In principle it should be much quicker to add the required gene to an embryo, and hence introduce it into the entire lineage, than to breed the animals by normal sexual means – especially in cattle, which do not breed at all until they are two years old and then have only one calf per year. Through genetic engineering, too, animals can be given genes that endow exotic qualities quite alien to the species – as indeed has been done with Tracy. In general, we can envisage some changes in animals that in principle seem relatively straightforward, and others that are bound to be much more difficult to bring to fruition. For example, a gene that is highly pleiotropic would be liable to affect the recipient in all kinds of unpredictable ways. So, too, would a gene that affected some fundamental system within the animal – so that early attempts to accelerate the growth of pigs by adding genes that produce growth hormone finished up producing deformities. Such endeavours must be abandoned on welfare grounds alone. Results are liable to be much more satisfactory if the transferred gene has a highly specific effect and does not affect the physiology of the animal as a whole. Fortunately, there are many such genes, including many that confer resistance to particular diseases – and we can expect that these will be the target for much future research in animal transgenesis, just as they are already for plants.

'Pharming' thus emerges as an outstanding outlet for genetic engineering. The kinds of genes that would be introduced would in general be highly specific in action, and should have little or no effect on the general wellbeing of the animal; but the products of those genes should be extremely valuable, justifying very great efforts to do the job well. When my colleagues at Roslin first began exploring genetic engineering in the early 1980s, pharming was not their only target. But it was certainly an important one.

The case for pharming

Biologists of all kinds make use of many hundreds of different kinds of complex chemical agents produced by living things, for all kinds of purposes. Antibiotics, mostly produced by fungi, are an obvious example. In this book we will encounter other equally complex molecules of various types used as laboratory agents, such as cytochalasin B (a material produced by a fungus). At least 120 proteins are known that are useful in human therapy.

Sometimes it is possible to synthesise such molecules in the laboratory but often it is not; and often it is far simpler and cheaper to produce complex biological agents within living cells. In general, at least so far, biotechnologists have preferred to produce arcane biological molecules within cultures of bacteria or fungi, for these organisms are relatively easy to work with and they raise no problems of welfare. Cultures of such microbes fitted with suitable genes can produce an ever-widening range of materials that they would not normally produce in nature. Thus insulin is a small and relatively simple protein molecule and microbes can produce the human form of it when fitted with the appropriate human gene. Insulin is, of course, used to treat diabetes.

But bacteria and fungi, wonderful though they are, have their limitations. They can produce some of the proteins employed in human medicine (when fitted with the appropriate genes) but others are beyond their abilities. Some protein molecules are simply too large for microbes to handle. Sometimes – often – the proteins need to be folded and generally modified after they are put together and microbes often lack the cellular machinery to do this. Thus many of the proteins that are required in human therapy have to be made in animal tissues. Some of course are simply harvested from human volunteers – who are the traditional source of blood products such as the clotting factors. Insulin has traditionally been obtained from pigs – for pig insulin is similar to human insulin. But it is now possible in principle to put human genes into animals, so that the animals produce the required proteins that can then be harvested. If the animals in question are mammals, and if they can be induced to secrete the required proteins in their milk, then harvesting becomes easy. This was the thinking behind Tracy.

Three among the many human proteins that might be produced by pharming are factor VIII, factor IX and the enzyme AAT. In healthy humans and other animals factors VIII and IX form part of a cascade of proteins that causes blood to clot when it is required to do so – generally speaking, when a wound exposes the blood to the air – but to remain fluid at other times, a balance that is very difficult to strike. In humans, deficiency of factor VIII leads to haemophilia A, while lack of factor IX leads to haemophilia B, or Christmas disease. The normal task of AAT is to counteract the effects of another enzyme, elastase, in the lungs. Elastase keeps the lungs flexible; but if it is not brought to a halt when its work is done, then it attacks the lung tissue. In cystic fibrosis and emphysema elastase gets out of control. Both conditions are distressingly common. Emphysema affects 100 000 people a year in Europe and North America alone. It is of course exacerbated by insults such as smoking, and may not occur without such insults, but the underlying cause is a genetic defect that leads to AAT deficiency. Cystic fibrosis is the commonest single-gene disorder among Caucasians: an astonishing one person in twenty carries the defective gene, and one in 1600 inherits this mutant gene in double dose and then suffers from the disease. Cystic fibrosis of course affects the pancreas as well as the lungs. Factor IX and AAT are already used therapeutically, but at present they have to be prepared from human blood plasma. But this is extremely expensive, it severely limits the supply, and it is potentially hazardous since plasma also carries the agents of disease such as hepatitis and AIDS.

This then, is the principle of pharming, at least as practised so far: add an appropriate gene to a zygote of a sheep (or some other milk-producing creature) and then, when the animal grows up, extract the gene product from its milk. There are, however, a few theoretical problems.

The problems with pharming

Problem one, before we even begin, is to decide *which* piece of DNA should be introduced. We know that each gene corresponds to a length of DNA, but only in some cases – perhaps only a few thousand out of the 60–80 000 genes in a human being – can we yet say which particular stretch of DNA corresponds to which gene. The new science of genomics is devoted to working out what genes each creature has, and where exactly they are positioned within the genome, and what their structure is. As I mentioned, my Roslin colleagues are heavily involved in genomics – in projects to map the complete genome of pigs, chickens, sheep, and cattle – the agricultural equivalents of the current, worldwide programme to map the human genome. But these are early days. Only a few genes are well enough known to be serious candidates for transfer.

The next problem – how to introduce the required stretch of DNA into a pronucleus – I have already outlined. Microinjection into a pronucleus has been the standard route up until now but although it is wonderful that such a technique works at all there are many difficulties. To begin with, of all embryos injected with foreign DNA, less than half incorporate it into their chromosomes; if the new DNA is not incorporated, then it soon disappears as the embryos develop. An awful lot of embryos are wasted, therefore. But even if the novel DNA is incorporated into a chromosome the incorporation may not be immediate. If it is incorporated straight away – before the zygote undergoes its first division – then copies of it should indeed appear in all the cells of the embryo as it develops, and all the cells of the animal that is eventually formed, including its eggs or sperm. But sometimes the injected DNA is not incorporated into the host genome until after the zygote has divided to form two cells, or indeed until it has divided further to form four or eight cells. In such cases the new DNA will appear in only some of the cells of the developing animal; an animal that has different DNA in different cells is said to be a *'mosaic'*. Mosaic animals commonly produce two different kinds of sperm or eggs: some containing the new DNA, and some not.

With present-day technology, too, the novel DNA is incorporated randomly into its new host DNA, and this can have untoward consequences. First, we have already seen that only five per cent of a DNA macromolecule actually consists of true genes that code for proteins; the rest is of unknown function (though it does affect *gene expression* overall), and some of this non-coding DNA occurs within the genes themselves, as introns. If the novel DNA happens to be incorporated into a stretch of non-coding DNA between functional genes, then it should not interrupt the function of those other genes. But if it is incorporated into a part of the host DNA that includes coding sequences, then clearly it can be highly disruptive.

Then again, only a small proportion – five or ten per cent is a reasonable estimate – of the genes are functioning in any one cell at any one time (or are expressed, as the technical term has it). The rest are shut down. (In short: only five per cent of the DNA in a cell actually functions as genes – meaning that they provide coding sequences for proteins; and only five to ten per cent of those genes are expressed at any one time – meaning that they actually make RNA which makes protein.) It is one thing to put a gene into a new cell and it is quite another to ensure that that gene is then expressed properly. It might well be introduced successfully but then remain permanently switched off. On the other hand, it could be damaging to the animal if a gene that was intended to express in the mammary gland also expressed itself in muscle or brain or what you will – especially if, for example, the product of that gene was a clotting factor. But then, in principle a protein like a clotting factor might be produced exclusively within the mammary gland but then leak into the rest of the body, and again cause trouble.

Of course, mechanisms exist within the genome to ensure that the creature's genes – all 60–80 000 of them in our case – are indeed switched on and switched off only in the right tissues at the right times; and the task of the genetic engineer is to integrate the transgene into the mechanisms that the body employs to control its own genes. At first sight this seems a very tall order indeed since the expression of each and every gene is influenced by a whole suite of mechanisms: messages from the cytoplasm, from other cells in the form of hormones, from the outside world in the form of daylight, or from hormone-like signals from other individuals, known as 'pheromones', and so on. Yet there is a way of achieving what is required. For, whatever the nature of the initial signal, the expression of all genes is ultimately mediated through a 'promoter region' of DNA that is attached to the gene. And it is possible (yet another serendipity!) to control the expression of a transgene by attaching it to a promoter that would normally control one of the animal's own genes.

This, then, in outline, is the scientific and technical background to pharming, and the reasons for doing it. So now we can look at what I and my Roslin and PPL colleagues have actually done, to turn the principle into reality.

But there are two distinct threads to this story. One leads directly to Tracy and to the current clinical trials of AAT, as outlined above. The other has been the attempt that I initiated in the mid 1980s to improve on the technique that produced Tracy; out of that attempt has come the modern technologies of cloning. The story may seem a bit messy – but that's because life is messy; and science is a slice of life.

The conventional side of the story: the road to Tracy

As I outlined in the last chapter, my colleagues at Roslin (though it was then called ABRO) began the work that led to Tracy around 1982 – the time of crisis, when the ruling ARC wanted to cut ABRO's finance by at least two-thirds. Such work requires two kinds of input. Clearly it needs molecular biologists who are able to identify appropriate genes (in this case human genes), and get them into the recipient animals (in this case sheep) in a form in which they will be expressed in the appropriate tissues, and at the appropriate time (which in this case means in the milk). But it also needs reproductive biologists who can supply zygotes to order, and either inject their pronuclei with DNA themselves or direct others to do it, and can then transfer the genetically transformed embryos into surrogate mothers, and bring them successfully to term. The molecular side of the work was initiated by John Bishop and Rick Lathe, and brought to an extremely successful conclusion by John Clark; and I provided the reproductive–embryological half of the team.

So I set up the protocols for obtaining embryos (we will discuss the kind of procedures involved in later chapters) and together with a postdoc called Paul Simons ('postdocs' are scientists with doctorates in their first professional jobs) we developed ways of injecting one of the pronuclei with DNA – although it is not at all easy even to see pronuclei in fresh zygotes. I did not do the injecting myself, however, since my hands shake. John Clark meanwhile developed the necessary molecular techniques and the first transgenic lamb was born at Roslin in June 1985.

This work – genetic transformation injecting DNA into one pronucleus of a zygote – continued through the late 1980s and led directly to Tracy, who was born in 1990 and is formally discussed in *Biotechnology* of 9 September 1991 in a joint paper by scientists from PPL and Roslin (though then it was still called the Edinburgh Research Station of IAPGR). To make Tracy, my PPL and Roslin colleagues first attached the gene for AAT to the promoter region of a gene that normally controls the expression of one of the proteins in sheep milk: to whit, the protein beta-lactoglobulin. Every cell in the sheep's body contains a copy of the gene for beta-lactoglobulin, together with its promoter, but it is only within the environment of the mammary gland, at times of lactation, that the promoter encounters the signals to which it can respond, and which induce it to switch on the beta-lactoglobulin gene. Thus the beta-lactoglobulin gene is expressed only in the mammary gland, and only during lactation. The same promoter attached to the AAT gene should ensure that this gene too is expressed only within the mammary gland, and only as the animal lactates.

Once they had attached the AAT gene to the ovine beta-lactoglobulin gene promoter, the scientists microinjected the resulting 'construct' into one of the pronuclei of zygotes from Scottish Blackface sheep (the local breed). They then grew the resulting embryos briefly *in vitro*, using techniques that had mainly been developed in human clinics for *in vitro* fertilisation. When the resulting embryos were a few days old, they then transferred the embryos into the wombs of sheep who had been primed with hormones so that they were ready to receive the embryos. The kinds of techniques involved are discussed in more detail later in the book, although as we will see it is more common in cloning work to raise the young embryos in the oviducts of temporary recipients, rather than *in vitro*.

These experiments produced five transgenic sheep, of which four were female. Among the females, predictably, the yield of human AAT in the milk was highly variable, although all four produced at least one gram of AAT per litre. But one of the transgenic sheep was outstanding. She began by producing more than 60 grams of the required protein per litre of milk and settled down to a steady 35 grams per litre. This is an astonishing yield. This was Tracy.

Tracy derives from research that began at ABRO but Tracy herself belongs to PPL. PPL produced more sheep like her, and their offspring have established commercial flocks of animals that all contain the human transgene, and in which all the ewes express AAT in their milk. The animals are simply multiplied by normal sexual means: transgenic rams and transgenic ewes between them breed transgenic offspring, which can then be cross-bred and selected to improve performance as in any other flock. The progeny of Tracy's peers now form two different flocks – one in Britain and one in New Zealand – and between them they have established PPL as a serious force in biotechnology. In 1987 PPL had only one employee but in September 1996 the company was launched on the London Stock Exchange and was immediately valued at £120 million.

Everyone involved in making transgenic animals by injecting zygotes acknowledges the shortcomings. In addition to those outlined above there is a more general point: that addition of DNA is the *only* form of genetic engineering that can be practised on zygotes. Biotechnologists who work with microbes, however – cultures of single-celled organisms – can not only add genes but also remove them ('gene knockout'), and this can be useful in many contexts. At least in theory, too, they can take out genes and alter them, and then put them back. Cells in cultures are easier than zygotes to transform for several different reasons. First they are laid out accessibly in a dish or flask – and it is possible to add genes to them simply by dropping DNA on the top, preferably perhaps enveloping the DNA in a fatty capsule to help it to penetrate the cells' outer membranes, which are also fatty. Then again, whereas a zygote is a one-off, cells can be multiplied in culture by the tens of thousands, or even the millions. Efficiency of genetic transformation is low – but so what? If only one cell in a hundred thousand is successfully transformed, that is good enough. Zygotes are intrinsically more difficult – yet efficiency clearly has to be much higher, for zygotes are precious commodities. Thirdly, zygotes are transient – racing to become embryos – while cells may live and multiply in culture for many weeks or months: plenty of time to make the necessary manipulations and indeed to monitor the results.

The only trouble is, of course, that if you transform animal cells in culture you produce – well, cultured transgenic cells. Of course it can be well worthwhile simply to transform cultured tissue in this way. For example, if a person has a disorder in which some specific tissue is defective – faulty pancreas cells, for example, leading to diabetes; or abnormal blood cells, leading to one or other of the inherited anaemias – then it should soon become possible to culture their cells, correct the defects, and return the repaired tissue to their bodies. But the aim of pharming, and the broader aim of genetic engineering in general, is not simply to transform specific tissues, but to make entire transgenic animals; and indeed if required to produce whole flocks, entire lineages, of transgenic animals. If transformed pancreatic tissue, say, is returned to an animal or a human patient it may help that patient to overcome a specific disorder. But only the pancreatic tissue will be transformed. Most of the animal or person will be entirely unaffected. Most importantly, the eggs or sperm of the tissue recipient will be entirely untouched; the recipient will *not* pass on the transgene to their offspring.

I was unhappy at first in my role in the genetic engineering project, providing zygotes; but I had an agreement with Roger Land that once the protocols for zygotes were set up, I would start looking for better ways. I perceived that we needed to culture animal cells, transform them *in vitro*, and then somehow derive whole animals from the cultured cells.

A way of doing this was already emerging from research else-where on mice, and in particular on cultured mouse cells known as *embryo stem cells* (or *ES cells*). We will say much more about

ES cells later in the book; but a few introductory comments are appropriate here.

The search for embryo stem cells

As I mentioned briefly in the last chapter, the first reports of ES cells date from about 1980. In essence, biologists found that they could culture mouse embryo cells in ways that retained their embryo-like status. This is highly unusual: cells change when they are cultured – they adopt specialised features that enable them to cope with the culture conditions (either that or they die). Embryo cells, when cultured, begin to adopt at least some of the specialised features that they would take on if they remained in the intact animal; that is, they begin to differentiate. ES cells also change somewhat in culture, but they retain embryonic features nevertheless.

We know that cultured ES cells retain embryonic features partly because they look somewhat embryonic, but mainly because they can be shown to *behave* like embryo cells. If, after culture, they are put back into young embryos – or at least, are put into the central mass of cells in the embryo known as the *inner cell mass* (or *ICM*) – then they will be incorporated into the embryo as if they had been there all along. The resulting embryo – containing its own original cells, plus the introduced cells – is then known as a *chimera*.

Once ensconced in the new, recipient embryo, the introduced ES cells are able to develop into *any* of the cell types that the inner mass cells would normally develop into: muscle, nerve, blood, what you will. Most importantly of all, however, ES cells introduced into a young embryo are perfectly capable of developing into germ tissue – which means they may be incorporated into ovaries or testes and so form eggs or sperm.

ES cells therefore seem to open a royal route to genetic engineering. The biologist can first take cells – ICM cells – from an embryo; then grow them in culture, where they develop into ES cells; and then subject the ES cells to genetic transformation

54

while they are being cultured. Then the transgenic ES cells can be transferred to another embryo to make a chimera: and sometimes some of those transformed cells will form part of the embryo's developing germ cells, and so in the end produce transgenic gametes, eggs or sperm. The transgenic animal can then by normal sexual means produce offspring that contain the introduced gene.

The method works beautifully in mice. But there is a snag. Cells that fulfil all the criteria of true ES cells – that is, cells that can be grown in culture and then behave like normal ICM cells when returned to an embryo – have been produced *only* from mice; in fact they have been produced only from one or a very few particular strains of mice. Embryo cells from sheep or cattle or pigs can certainly be cultured, but then they differentiate too much; at least, they do not retain the ability to develop into any tissue when put back into embryos.

To me in the early 1980s, however, ES cells seemed very promising indeed. I felt that if we could develop ES lines from sheep, they could then be genetically transformed in culture, then used to make sheep chimera embryos, and so go on to give rise to lineages of transgenic sheep. So, with Roger Land's blessing, I began the search.

Keith is 10 years younger than I am – he was born in 1954 – and while I was establishing my new line of research at ABRO in the early 1980s Keith was working on his 'first PhD' at the Marie Curie Memorial Foundation, which is a cancer research institute in Surrey. There he explored a range of everyday materials that could damage chromosomes and so (in some cases) trigger cancer. He realised then that cells in cancers are peculiarly flexible. They may begin as specialised, differentiated cells – and then their nature changes completely as the cancer progresses. So, as he says, 'If you open up a tumour you find all sorts of tissues in there! Hair, nails, bits of bone, fat, muscle – all sorts of things! So obviously the supposed programming of the cells has gone completely haywire.' In other words, while I was struggling with ES cells, which is essentially an exercise in cell differentiation, Keith was beginning to get involved in the problems of differentiation from a quite different angle.

In truth, the work on ES cells, logical though it seems, never quite worked out. Along the way, though – and crucially to our story – it gave rise to the technology of cloning by nuclear transfer. But then, when the first forays into nuclear transfer had been made, Keith joined me at Roslin (then called IAPGR) in 1991 and contributed his own ideas on cell culture, and on cell programming and reprogramming. Out of this collaboration came the technologies that led first to Megan and Morag, and then to Dolly and the rest. The search for ES cells, which started the whole process, came in the end to seem largely irrelevant. None of the cloned sheep at Roslin (or any other cloned livestock anywhere else in the world at the time of writing – July 1999) have been developed from ES cells – although I still think that it might be advantageous to clone from ES cells, if only we could produce them.

But I am beginning to run ahead. Before we look at Keith's and my ideas in more detail, and at why and how the cloning technologies of Megan and Morag and Dolly (and finally Polly) emerged from our combined efforts, we should take a few chapters to establish a few more background ideas. Again, we might as well begin at the beginning with the somewhat deceptively simple question: 'What is a clone?'

PART II

The Science of Cloning

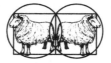

CHAPTER 3

So What Exactly is a Clone?

Ian Wilmut and Keith Campbell:

We do not seek simply to clone animals – to produce facsimiles of existing creatures. We are not concerned primarily to multiply elite livestock, and still less do we want to clone human beings. This was never on our agenda; it is just what other people thought was important. Cloning for us is and has always been an exercise in science – finding out how cells work – and a technology that enables the genetic transformation of animals. Cloning, neverthe-less, is what we do, so we should ask what cloning is and what it is not. The details matter. They are at the heart of the science and also, in practice, the facts of the case do bear upon ethical decisions and theological attitudes. The word 'clone' has many connotations, and is used to describe several different (or at least clearly distinguishable) biological entities, and we should look at the range. It is at least pertinent, for example, that Dolly is not an absolute, hundred per cent replica of the old ewe who provided her first cell (who we might call her 'clone mother'). She is not as similar to her clone mother as two identical twins would be. She is merely a genomic, or a DNA clone. That might not be a huge fact, but it is a fact nevertheless. So let us look at the underlying biological concepts.

All living things reproduce; reproduction, or 'replication', is one of the distinguishing features of life. The easiest way to reproduce is simply to divide. This is the way DNA replicates itself (or at least is replicated with the aid of enzymes): the two strands of the double helix split apart and each single strand then makes a complement of itself. Bacteria and protozoans multiply simply by dividing in two: a form of replication known as '*binary fission*'. All the cells of the body reproduce themselves by such division (although some of them stop dividing once they have reached their final, specialised stage). Flowering plants in general reproduce by seeds which for the most part are the products of sex, but most are also able to reproduce themselves by some variation on a theme of budding – by suckers, stolons, rhizomes, tubers, or what you will. Reproduction by such conceptually straightforward means is said to be '*asexual*' – meaning 'without sex'. Cloning is a form of asexual reproduction; although actually it is not just one form of asexual reproduction, since there are many routes to cloning.

The term 'clone' may be either a noun or a verb. As a noun it is used in two ways. It is applied first to the individual offspring that are produced by asexual reproduction, so that two Cox's orange pippin trees, for example, each produced by cuttings, are said to be 'clones' of each other. But it is also applied to all the asexually produced offspring collectively, and to their identical parent; so that all the Cox's orange trees there have ever been, going right back to the very first one that grew from a pip in the 19th century, collectively form a clone. Many species of grass reproduce by runners or stolons (as well as by sexual seeds), and then an entire patch of sward may be a clone. As a verb, 'clone' is used both intransitively and transitively – that is, plants that reproduce asexually are said to 'clone'; and a scientist who induces asexual reproduction in some creature is also said to 'clone' it. Thus elm trees clone themselves to form entire copses, and we cloned Dolly from cultured mammary gland cells.

Human beings are, of course, species chauvinists. We tend to think that we represent normality, and judge all other creatures accordingly. Human beings do not – at least in a state of nature – reproduce asexually. For us, sex is the necessary preliminary to replication. To us humans, asexuality has a kind of feral feel to it. But human beings are not alone on this planet. There are many other ways of being a living creature.

In general, living things can be very broadly divided into those that keep their DNA neatly enclosed within a nucleus – and these are called '*eukaryotes*' – and those in which the DNA is not so neatly enclosed, but is simply packaged in various ways within the body of the cell – and these are called '*prokaryotes*'. Eukaryotes include protozoans, seaweeds, plants, fungi and animals; prokaryotes include the creatures commonly known as 'microbes', which are the bacteria and the archaea.

Most groups of creatures, both prokaryotic and eukaryotic, practice both asexual *and* sexual reproduction. Among prokaryotes, asexual reproduction – cloning – is the *normal* way to replicate, although bacteria do practise various forms of rudimentary sex as well. All eukaryotes must at one time have been sexual – it is impossible to see how they could have evolved otherwise – but many have now abandoned sex altogether. Asexuality is again common in *all* the major eukaryote groups. Protozoans, like *Amoeba*, practise binary fission. Seaweeds and fungi generally reproduce sexually by spores but many fungi also reproduce simply by spreading their hyphae (filaments), and also by producing asexual spores: spores produced without the preliminary benefit of cell fusion. There are many fungi (quaintly known as *fungi imperfecti*) in which sexual reproduction has never been observed. Most plants, too, are both sexual and asexual: mosses and ferns spread asexually and also produce sexual spores; flowering plants reproduce by seeds but also clone themselves asexually, while gardeners clone many garden plants by cuttings. Many animals reproduce asexually as well, at least some of the time: corals, many insects, and a range of vertebrates including some fish and lizards and even, under artificial conditions, some birds. All animals must be descended from sexually reproducing forms, but, as mentioned above, some have abandoned sex altogether.

In fact asexual and sexual reproduction are very different kinds of processes, each serving a different evolutionary role. Asexual reproduction is the obvious way to replicate: start with one individual, split down the middle, and then you have two, or many. By contrast sexual reproduction is a bizarre and even a perverse way to replicate. Two protozoans that seek to reproduce sexually must first fuse, to produce one. Multicellular creatures like oak trees and human beings produce specialist sex cells known as *gametes* – and again, two gametes must fuse to form just one new embryo. Sex, in short, is *anti*-replication. Replication implies that one individual divides to become two or more. But with sex, two combine into one.

Furthermore, although sedentary creatures like mussels and oak trees are content to scatter their gametes (often only the male gametes) to the tides or the winds, creatures that are not sedentary, including snails and bees, sheep, and human beings, must actively seek out their mates. Mating is an extremely tense, difficult, time-consuming, and often very dangerous pursuit: ask any male spider or mantis, risking death by cannibalism, or any stag or lion or elephant bull, risking death from mauling by the antlers, teeth, and sheer bulk of rivals (and *vide* all literature). Yet reproduction by sex is innately inefficient, for we cannot escape the arithmetic: two cells fuse to become one. For creatures who generally produce only one offspring at a time, like human beings, orang utans and giant pandas, this means that two entire individuals must put their lives at risk just to produce one offspring between them. It is only because we are such human chauvinists, and think that what we do is by definition the 'norm', that we fail to find this odd. It took a great biologist – John Maynard Smith – to point out that if a human being, or whale, or codfish, or oak tree ever evolved that could produce *asexually*, without needing a mate, then that creature would soon overwhelm its sexual rivals. Head for head, such a lineage would leave twice as many offspring.

The puzzle, in short, is to explain why sex happens: why, despite its obvious arithmetical and social drawbacks, it is so common in nature – probably as common as asexual reproduction. The answer is that sex has no direct, conceptual connection with replication at all. Sex is not about multiplication. It is about the mixing of genes from different organisms. The question arises, of course, as to *why* different organisms should want to mix their genes, and to this there are various answers. The old-fashioned answer is – or was – that by mixing their genes creatures produce an effectively infinite range of new genetic variations, and this enables evolution to proceed much faster than it otherwise would. For example, if one individual in a population undergoes a genetic mutation that happens to be beneficial – call it mutation A – and another undergoes a different beneficial mutation – B – then, by combining, they can produce offspring that possess A *and* B, which would be a kind of super-creature. Given that all beneficial mutations are rare, it would take a virtual infinity of time for any one individual, without the benefit of such mergers, to develop *both* A *and* B. Tempting though such explanations are, however, they cannot explain why sex evolved in the first place. Such genetic swapping does indeed produce evolutionary benefit – but the benefits are long term. As John Maynard Smith pointed out, creatures that required sex for reproduction would still lose out *in the short term* to creatures that merely replicated themselves asexually, and so produced twice as many offspring per head in a given time. Of course, long-term evolution is wonderful, and without it there would be no human beings or elephants or oak trees. But the long term never comes into being if it is swamped in the short term. Evolutionary change is a serendipity; a useful side-benefit of sex. But it cannot be the driving force behind it.

The task, in short, is to explain the short-term benefits of genetic mixing, and here there are two main kinds of proposals. One says that it is highly beneficial in the short term to produce offspring that resemble the parents (of course) but are also slightly different. Professor Bill Hamilton of Oxford University has suggested that such variation provides short-term benefit against parasites, which would rapidly adapt to creatures that remained completely uniform from generation to generation, but find it hard to get a foothold when the lineage constantly varies. A totally different kind of explanation hinges on the fact that in any one generation there is liable to be a great deal of mutation, but that most mutations are harmful. In sexual reproduction an individual passes only half of his or her genes to each gamete, and some at least of those gametes will be relatively free of harmful mutations. Only the good gametes will survive – those lacking mutations that are too harmful – and by fusing they provide a new generation of offspring that is relatively free of mutations. This idea has been around for a few decades and seems to be growing in popularity.

In short, sex did *not* evolve primarily as a means of replication. It evolved as a means of mixing genes from different individuals – and there do seem to be short-term as well as long-term benefits in this. But it is easy to convert sex into a means of replication. After all, a multicellular creature can produce many gametes, and once these are fused they do indeed form new individuals; and that is clearly a form of multiplication, albeit an arithmetically inefficient one.

In fact, despite the theoretical drawbacks of sex as a means of replication, the gametes that have evolved primarily as a means of mixing genes have also, many times, *re*-evolved as an apparatus for *a*-sexual reproduction. Often, indeed, the egg simply begins to develop into a new individual without fertilisation by a sperm. This form of cloning is called *parthenogenesis* (from the Greek 'parthenos' meaning virgin; hence, 'virgin birth').

In practice, parthenogenesis takes several forms. As discussed more fully in chapter 5, the body cells of animals and plants typically contain two complete sets of chromosomes (one inherited from the mother and one from the father) and are said to be *diploid*. But the gametes, in their mature form, contain only one set of chromosomes and are then said to be *haploid*. Diploidy is restored when two haploid gametes fuse. Parthenogenesis invariably involves eggs, not sperm, for sperm are highly reduced cells – just a bag of DNA with a tail for mobility – and they do not have the wherewithal to develop into new embryos. Haploid eggs arise from diploid germ cells, which are known as *oocytes*. (Confusingly, oocytes are also commonly referred to as 'eggs' but we need not let that delay us here.)

Sometimes parthenogenesis involves development directly from diploid oocytes and sometimes it involves development from haploid eggs. Thus, dandelions and related plants such as hawkweeds reproduce parthenogenetically – and have indeed abandoned sex altogether – but all of them are diploid; so parthenogenesis begins with the oocyte. But in bees, wasps and ants, the queens produce diploid females by normal sexual means – these become the workers and new queens – but produce the males (called drones in the case of bees) by parthenogenesis from haploid eggs. Thus male ants and bees are themselves haploid. Note that when a diploid female produces diploid offspring parthenogenetically then those offspring are genetically identical to each other and to her, and the offspring and the mother collectively form a clone. But when a diploid female produces haploid offspring parthenogenetically then those offspring are not identical to each other, or to the mother. Each gamete, from which each haploid offspring develops, inherits only fifty per cent of the mother's genes. But the different gametes are all different: they do not each inherit the *same* fifty per cent. We cannot, then, say that asexual reproduction *is* cloning. Asexual reproduction *usually* implies cloning, in some form or other. But here is a form of asexual reproduction that does not lead to cloning.

Parthenogenesis is rare among vertebrates but by no means unknown. Some lizards reproduce parthenogenetically – beginning, like dandelions, with the diploid oocyte so that they are all diploid – and some indeed have abandoned sex altogether. Some fish can also reproduce parthenogenetically. Modern breeders are able to induce fish such as salmon to reproduce parthenogenetically; and parthenogenesis has even been reported in domestic turkeys. In mammals, however, parthenogenesis seems to be a non-starter. As discussed more fully in chapter 6, mammals exhibit the phenomenon of *genomic imprinting* – meaning that there are *some* genes which behave differently when inherited from the mother to when inherited from the father. As a result of this, mammalian embryos will not develop properly unless they inherit genes both from a father and from a mother. The extreme feminist dream of all-woman reproduction does not seem to be possible by parthenogenesis – although some version of it might be achieved by other means (that is, by variations on a theme of cloning by nuclear transfer; more of this later).

However, mammals do practise asexual replication in a minor way; sometimes it seems 'by accident' but, in one case at least, as a matter of course. Thus an embryo sometimes divides in the womb to produce identical twins. Identical twins together form a clone; each is a clone of the other (though they are *not*, in this case, clones of the parent!). In nine-banded armadillos the embryos invariably divide when they reach the four-cell stage, to produce a set of identical quads. Why nine-banded armadillos have evolved to give birth to identical quads has not been convincingly explained. Not everything that happens in nature can sensibly be seen as an adaptation that truly enhances survival. Nature is quirky. The four-strong clones that armadillos produce are probably just one of those things. It surely would be better for them to produce four slightly different offspring, each with its own particular strengths. On the other hand, armadillos have been around for a long time so they must be doing something right (and they are among the few wild mammals in North America that are actually increasing their range).

Thus if we look at the broad sweep of nature, we find that asexual reproduction is extremely common – probably as common as sexual reproduction – and that reproduction without sex usually, though not invariably, means cloning. Cloning happens to be unusual among vertebrates, however, and is extremely unusual among mammals – the class of creatures that includes human beings. One of the common routes to cloning – parthenogenesis – seems to be closed to mammals because of the specific phenomenon of genomic imprinting. These, then, are the broad facts: cloning is a very natural thing (meaning that it occurs throughout nature) but it is not, except in rare instances, a mammalian thing. These background ideas do not bear directly on the issue of whether the artificial cloning of mammals is right or wrong but they do perhaps suggest that cloning is not quite so outlandish as sometimes presented. It's just that mammals usually don't do it. Mammals have gone down another evolutionary route.

We began this chapter by noting that bacteria and protozoans can clone themselves simply by dividing in half. When multicellular creatures clone, however, then this implies that a whole new individual – commonly with billions of cells – arises from just one or a few cells produced by the parent individual. The division and redivision of a single cell to produce a multicellular creature raises a whole new set of biological issues that are crucial to the kinds of technologies that have produced Dolly.

The cloning of multicellular creatures: totipotency, pluripotency and differentiation

Creatures like us do not generally replicate by cloning but the development of all multicellular creatures – including us – *involves* cloning. That is, each of us began life as a single-celled embryo – a zygote, produced initially by the fusion of egg with sperm. Then the zygote divided, and divided, and divided again by a process of cell division very similar to the asexual reproduction practised by protozoans. Each of our body cells, then, is a clone of all the other cells, and of the original zygote. All the cells of our body collectively form a clone.

But although the cells of the body all arose from one single cell, and so are all clones of that cell, they are extremely varied. Nerve cells are nothing like muscle cells, which are nothing like skin cells, which are nothing like the white blood cells of the immune system, and so on. There are many billions of cells in an adult human body but also there are many different *kinds* of cells. How come, if all were cloned from the same original zygote?

Two phenomena are implicated: first, the issue of *totipotency* and its variant *pluripotency*; and secondly, that of *differentiation*. Thus, a zygote is clearly capable of giving rise to *all* the different kinds of cells in the body. It is, accordingly, said to be 'totipotent'; the etymology of the word is obvious. By contrast, when skin cells divide they merely produce new skin cells; and ultra-specialist cells such as those of the nerves and muscles do not divide at all. Perhaps the most specialised of all are the red blood cells of mammals, which have lost their nuclei; and without their nuclei they obviously cannot divide. (Note in passing, though, that the red blood cells of reptiles and birds retain their nuclei, although they still do not divide.) Between the extremes of totipotency and extreme specialisation is the state of 'pluripotency'. Thus the stem cells of the blood and some other tissues are able to give rise to the various forms of white and red blood cells, but they cannot give rise to, say, nerve or muscle cells. Such stem cells are said to be 'pluripotent'; and there are many examples of pluripotent stem cells around the body, renewing tissues of the kind that are particularly prone to wear and tear. The 'embryo stem (ES) cells' that we met in the last chapter are a special kind of stem cell. Derived from the inner cell mass (ICM) cells of young embryos, they apparently retain a very high measure of pluripotency; that is, under suitable conditions, they are able to give rise to cells that can differentiate into any of the tissue types that ICM would normally give rise to.

Differentiation is the name given to the process by which some of the daughter cells of the zygote divide to produce muscle cells, and others to produce liver cells, and so on. The mechanism of differentiation is now understood to some extent – although there is a great deal left to learn, and the understanding so far has been hard won. The concept of differentiation is a key theme of our work, and we had best discuss it as the book unfolds. Crucially, though, you can see intuitively that totipotency and differentiation seem to be at odds with each other. The more differentiated a cell becomes – the more it is committed to be skin, or nerves, or whatever – the less it seems liable to be totipotent. This commonsense observation dominated attempts to clone mammals artificially – and the great innovation that lies behind our success with Dolly was to show that, in this instance at least, common sense and intuition cannot be relied upon. Keith has long argued that differentiated cells – including differentiated mammal cells – *can* be reprogrammed, and can be persuaded to behave again as if they were totipotent embryo cells. But we will discuss all this at length in chapter 8 and the ensuing chapters. This insight – a radical departure from tradition – was and is the key to the last, successful phase of our work.

With these general biological notions in mind we can look briefly at the means by which scientists have induced various creatures to clone that would not normally do so, or in ways they would not normally adopt; or have simply turned natural cloning to their own uses.

Artificial cloning

Gardeners often propagate plants simply by breaking off twigs and sticking them in the ground (or, at least, this is the essence of it). All the 'varieties' of fruit that grow on bushes and trees – apples, grapes, raspberries – are in fact clones, multiplied by stem cuttings. Many other plants are multiplied from pieces of root.

The fact that a cutting will often grow into a whole plant implies that some at least of its cells are totipotent, or at least highly pluripotent. If a piece of stem is stuck in the ground, then some of the cells differentiate to give rise to roots; and cells in a root cutting may differentiate to produce a new stem, which in turn produces leaves, and flowers, and then seeds – all the multifarious cells of which the plant is composed. Some animals have a comparable talent, and thus an entire new individual may be generated from a (large) fragment of starfish. In general, though, plants seem more inclined to retain totipotent cells throughout their lives than animals do. You obviously cannot regenerate an entire human being from an ear or a piece of leg, except in Greek mythology.

Typically, when plants are wounded, they close the wound by generating 'callus' cells – which are relatively undifferentiated, and retain or recover some totipotency. It is from callus cells that new specialist tissue arises – stem or root, whatever is lacking. Biologists attempted at the beginning of the 20th century to generate whole plants from callus cells in culture but failed more or less entirely until the 1930s when the first plant hormones were discovered; then, by adding these hormones, it became possible to multiply callus cells in culture, at least for a time. Further work to adjust the ratio of different hormones finally produced success in 1958 when an American scientist generated an entire carrot from cultured carrot callus cells. The first plant clones to be generated *from single cells* in culture were orchids, in France in 1960.

Cloning from culture is now a common technique in experimental horticulture (and animal cloners might pick up on some of their ideas). Cloning from cultured cells is of enormous commercial value for propagating coconut and oil palms – plants that do not naturally reproduce asexually at all, and cannot be multiplied by cuttings like apples and raspberries.

But the cloning of animals has in general been less simple, and has lagged considerably behind that of plants, largely because animal tissue is so much less inclined to retain any totipotent cells. The scar tissue with which animals seal their wounds is not directly comparable to the callus tissue of plants. We will require much of the rest of this book to describe the long and tortuous journey that has finally produced Dolly – a journey involving several generations of scientists, through well over 100 years. But a rapid overview is in order, to make the detailed account easier to follow.

Conceptually, the simplest way to clone an animal is simply to split an embryo: imitating what nature sometimes does *in utero*, to produce identical twins. Clearly each cell of a two-celled embryo *is* totipotent. But it is also possible to split young embryos artificially, and again produce whole animals from the individual cells. The number of cloned individuals that can thus be produced varies from species to species, and is severely limited because the cells of successive generations are smaller and smaller and are soon too small to develop. Thus mouse embryos can be split at the two-cell stage to produce identical twins, while cattle and sheep embryos can be split at the eight-cell stage, although no one has ever produced more than five of the theoretically possible octuplets. Survival can be enhanced, however, by inserting the cells from an eight-cell embryo – or indeed from an older embryo – into another embryo to produce a chimera. Chimeras that incorporate ES cells – cells that have been multiplied in culture – are a special example of this.

The great advance, however, has been not simply to split embryos, or to make such chimeras, but to transfer the nucleus of an embryo cell into the cytoplasm of an oocyte or zygote whose own genetic material has already been removed (when they are said to be 'enucleated'). As we will see in the next chapter, progress towards this kind of technique began in the 19th century but was not finally achieved until the 1950s, when it was carried out in frogs. In mammals, cloning by 'nuclear transfer' was first fully successful – that is, produced live offspring – in the mid 1980s: not in mice, as might have been anticipated, but in sheep.

Note, now – a crucial point – that clones produced by nuclear transfer are *not* directly comparable to clones produced by embryo splitting, whether these are produced artificially or naturally, in the form of identical twins. To appreciate this, we should divert slightly, to look briefly at the structure of the cell.

The nature of the cell – and why Dolly is not a 'true' clone

We have said that prokaryotes like bacteria have simple cells, with DNA packaged in various ways throughout the cell body; while protozoans, slime moulds, seaweeds, fungi, plants, and animals are 'eukaryotes', and keep their DNA pleasantly cocooned within a discrete nucleus, surrounded by its own nuclear membrane. In fact, the differences between prokaryotic and eukaryotic cells run much more deeply than this. Eukaryotic cells – the kinds that human beings have, and sheep, and oak trees – are wonderfully complex. If each prokaryote cell is compared to a house, then each eukaryote cell is like a city.

Through the light microscope, the 'standard' eukaryotic cell looks like a fried egg – albeit superficially, and only when suitably stained. The 'yolk' is called the *nucleus* and the 'white' is called the *cytoplasm*. In the bad old days the cytoplasm was loosely referred to as 'protoplasm', as if it was just some structureless jelly, a pool in which the nucleus could wallow. In truth, however, the cytoplasm is enormously intricate and tightly organised; it's just that its intricacies are not so easy to see with a simple light microscope and so have become apparent only as microscopy and the techniques of biochemistry have grown in stature.

The prime task of the nucleus is to provide an environment in which the DNA, coiled into its chromosomes, can function most freely. The nucleus is surrounded by a membrane layer, sometimes simply called the '*nuclear membrane*' and sometimes the '*nuclear envelope*' (abbreviated to 'NE'). The latter term is appropriate because the envelope is in fact formed as a fold in the 'cytoplasmic reticulum' (of which more later). The nuclear membrane is able to maintain a DNA-friendly environment inside, because it has evolved to allow only particular, selected materials to enter and to leave. As we will see in later chapters, however, the membrane disappears as cells are preparing to divide, and reappears after the division; and when it is absent various 'factors' are able to flow in from the surrounding cytoplasm, and interact with the chromatin and so with the DNA, which is the key component of the chromatin. (*Chromatin* is the general term for all the materials of which the chromosomes are made). Thus the membrane always acts as a gate-keeper, but it also functions as lock-keeper; at critical times, it stands aside altogether, and allows the floods to enter. Within the nucleus is an extra dark-staining body known as the *nucleolus*, whose task it is to construct the ribosomes.

The cytoplasm, which looks like structureless egg-white under the ordinary light microscope, is in fact held together by an interfolding series of membranes known collectively as the '*endoplasmic reticulum*' (often abbreviated to 'ER') and the '*cytoskeleton*'. The cytoskeleton is a framework of fibres of two different types that has many functions. First, like any skeleton, it keeps the whole cell together: prevents it from being structureless jelly. The various organelles of which we will speak – ribosomes, mitochondria, etc. – are held in position. Folds in the reticulum form the NE, and also envelop the mitochondria. The membrane also largely controls the flow of messages throughout the cell, holding on to some agents, allowing others to pass through. As we will see, it is necessary to soften the cytoskeleton with a drug (cytochalasin B) when removing or adding nuclei during the cloning process.

An exhaustive list of all the various structures ('organelles' and 'plastids') that populate the cytoplasm and contribute to its overall function would occupy an entire textbook, but four in particular must be mentioned. The *ribosomes* we have met in chapter 2: small bodies containing RNA where proteins are assembled by joining amino acids together, according to instructions issued originally by the DNA in the nucleus. The '*Golgi apparatus*' is a series of folds in the endoplasmic reticulum which under a suitable light microscope looks like a pile of dinner plates, and is to the individual cell what the liver is to the whole body: a kind of processing plant. The *mitochondria*, dotted throughout the cytoplasm, are commonly called 'the power houses' of the cell. Their task is to process nutrient, and provide energy. 'Power house' is a somewhat macho metaphor; they might as lief be compared to *pâtisseries*. The oocyte is wonderfully rich in mitochondria – it must provide the energy for the first exertions of the developing embryo – and might in this metaphor be compared to a Brussels side-street, where every second shop sells pastries. Finally, many of the cells of plants, though not of animals, contain *chloroplasts*, replete with the pigment chlorophyll, which converts solar energy into chemical energy.

At the start of the 20th century a Russian biologist, C. Mereschkowsky, proposed that the intricate structure of the eukaryotic cell must have arisen in large part as a coalition of various prokaryotes. In particular, the mitochondria and chloroplasts could be seen as prokaryotic would-be parasites that had found lodging within the body of some larger prokaryote, and stayed there – eventually evolving into the organelles we see today. Like many great and revolutionary ideas in science this one was more or less ignored. But in recent decades the American biologist Lynn Margulis became its champion, and has elaborated upon it.

Now this coalition idea is effectively the orthodoxy. Among other things, it explains a phenomenon that otherwise would make no sense at all: why it is that both mitochondria and chloroplasts contain their own DNA. In fact, about three per cent of the total DNA in an average cell is mitochondrial. The structure of mitochondrial DNA is more like that of a bacterium than it is like that of the DNA in the eukaryotic nucleus. Clearly, in the two billion years or so since the coalition first took place, mitochondria have lost many of their original genes, so they are no longer capable of free living. Some of the genes that were orginally mitochondrial have clearly migrated to the nucleus. The mitochondrial DNA certainly contributes to the life of the cell, mainly through the effect on the cell's energy supply, but also in other ways: and some inherited diseases are known that are caused by mutations of the mitochondrial DNA. In general, the mitochondrial DNA is controlled by the nuclear DNA. The two sources of genes work in close collaboration.

With these ideas on board, we can see that there is a profound difference between the kinds of clones that are produced by nuclear transfer – like Dolly – and the kinds produced by embryo splitting, as in the natural generation of identical twins. For identical twins have identical DNA and also have identical cytoplasm; the cytoplasm of the original embryo cell simply splits, after the DNA has duplicated. But Dolly and her kind were made by transferring a nucleus from a cell into the cytoplasm of an egg cell that comes from a different animal. So Dolly and the ewe who provided the original nucleus have identical DNA but they do *not* have identical cytoplasm. In fact Dolly is not a 'true' clone of the original ewe. She is simply a 'DNA clone' or a 'genomic clone'.

We should not get too carried away by this. The ewe that supplied the original nucleus for Dolly was a Finn-Dorset, and the ewe who supplied the cytoplasm was a Scottish Blackface – a very different breed. But Dolly herself is very clearly a Finn-Dorset. No show judge would ever suggest any different. The nuclear DNA undoubtedly prevails. On the other hand, DNA does not operate in isolation. It is in constant dialogue with its cytoplasmic environment; the cytoplasm makes a difference. The mitochondria clearly can make a particular difference, since they contain their own DNA. How would a cow made by transferring a Friesian nucleus into Hereford cytoplasm differ from one made by transferring an identical nucleus into, say, an Ayrshire's cytoplasm? As yet nobody knows; but it would certainly be worth finding out, both for commercial and for theoretical reasons. Note, too, the point made earlier, that the four young rams we produced at the same time as Dolly – Cyril, Cecil, Cedric and Tuppence – are very different creatures, even though they are nuclear clones; and perhaps (no one knows if this is the case) their differing mitochondrial genes (or other cytoplasmic factors) account for some of the variation.

However, as discussed in later chapters, we make embryos by

fusing the *whole* cell that donates the nucleus to the cytoplasm of an enucleated oocyte. In common with other modern cloning scientists we do *not* first remove the nucleus from its own cytoplasm before transferring it. So, in fact, although the cells in Dolly's body are descended from a cell which mainly contained Scottish Blackface cytoplasm, that cell also contained some Finn-Dorset cytoplasm surrounding the donor nucleus. To be sure, the oocyte that came from the Scottish Blackface is far, far bigger than the cultured body cell that supplied the Finn-Dorset nucleus, so we might suppose that the meagre Finn-Dorset contribution would simply be swamped. In fact this does seem to be the case: mitochondria introduced into a new host cell do not seem to fare as well as the residents. But such things should not be taken for granted.

Finally, it would in principle be possible to make true clones even by nuclear transfer. We might, for example, transfer a body cell from a ewe into oocyte cytoplasm from the same ewe. That way, the offspring would contain a clone both of the 'clone mother's' DNA *and* of her cytoplasm. This would be a form of artificial mammalian parthenogenesis – a method that could, if applied to humans, realise the ultra-feminist goal of the self-replicating female (although the 'self' would have to be aided by attendant scientists and clinicians). These are niceties; but they could be significant niceties.

But we have made a further conceptual leap beyond that of mere nuclear transfer – and that is to multiply the donor cells in culture before transferring them (or their nuclei) into enucleated oocytes. Then, as already intimated, we have leapt even further, and found ways of *restoring* totipotency to cell lines that once would have seemed to be differentiated beyond recall. These two steps distinguish our work from all that has gone before: multiplying the cells in culture before transferring their nuclei to make new embryos; and creating new embryos from cells that are already differentiated, by reprogramming their genomes. The theory that lies behind this endeavour is described in chapter 8; and the method, which produced Megan and Morag and then Dolly, is outlined in chapters 9 and 10.

But this chapter is making cloning sound easy. To appreciate what has really been involved we should begin the story of animal cloning from its beginning. That was way back in the 19th century. Big ideas in science take a very long time to unfold.

CHAPTER 4

Embryos and Clones: Early Days

Ian Wilmut and Keith Campbell:

Because science seems to progress so rapidly, people tend to assume that every idea that is worth taking seriously must be hot from the presses and that last year's notions must be dead meat. In truth, the great ideas of science take decades, even millennia, to develop. Scientists in all fields constantly revisit discussions that took place in the 19th century, or the 18th – or even in classical times. Science in truth is a deeply historical, inescapably collective pursuit that has unfolded throughout human history.

Thus we can trace the sciences that eventually enabled us to produce Dolly back to the 17th century, when modern science as a whole is generally acknowledged to have begun. Natural philosophers perceived that the Universe runs by 'laws' which could be discovered by direct observation of nature, and experiment. By this means Galileo and Newton gave rise to recognisably modern physics while traditional naturalists and apothecaries began their own transformation into true biologists; mysticism began to fall away as they came to see living systems as explicable *mechanisms*, controlled in the end by the same laws that regulated the stars and planets. William Harvey (1578–1657) observed the motions of the blood, then interfered by a logical series of interventions with its flow, and hence deduced the function of the heart and the nature of the circulation – a model of scientific method that is unsurpassable. Technologies emerged to aid investigation. The telescope transformed astronomy and hence, more broadly, physics; and the telescope begat the microscope, which transformed biology. With the microscope biologists could *see* directly the underlying mechanisms of life, or at least began to think that this was possible.

Two of the greatest of the early microscopists came close to discovering that the bodies of living creatures are compounded from a huge assemblage of cells, though neither quite did so. In Italy Marcello Malphighi (1628–94) showed, among other things, the existence of the capillaries, the fine blood vessels in the body tissues that run between the arteries and the veins – the discovery that rounded off Harvey's vision of the circulation. At about the same time a Dutch linen-draper called Anton van Leeuwenhoek (1632–1723) made his own microscopes – hundreds of them, in the course of his long life – and discovered a host of creatures too small to see with the unaided eye which he called 'little animalcules'. Nowadays we call such animalcules 'microbes' and know that they include very various creatures: 'bacteria' and 'protozoa' or 'protists'. Each of them – we would say in modern parlance – consists of a single cell. Peter the Great was among the many celebrities who flocked to Delft to see the little animalcules for themselves. Microscopes caught the imagination, as well they might.

Yet it was not until the 19th century that microscopists perceived that the bodies of all large creatures are built from cells, just as the mightiest cathedrals are compounded from stones. In 1838 the German botanist Matthias Schleiden (1804–81) proposed that every kind of plant, however big it may be, is put together from a myriad cells, or materials produced by cells; and the following year the German physiologist Theodor Schwann (1810–82) suggested in *Mikroskopische Untersuchungen (Microscopical Researches)* that 'cellular formation' might be 'a universal principle for the formation of organic substances'. Schwann proposed that the individual cells simply condensed out of 'nutrient liquid', like crystals of salt precipitating out of a rock pool, but this misconception was soon put to rights. In 1841 the Polish-German anatomist Robert Remak (1815–65) described how cells divide and in 1855 Rudolf Carl Virchow (1821–1902), a German pathologist, proposed the dogma *omnis cellula e cellula* – 'all cells come from cells'. In other words, the vast battalions of cells of which a mouse or a sheep or a human being is composed (or an oak tree or a moss or a mushroom) each derive from a single, initial cell, dividing and redividing. (Note in passing how prominent was the German role in this phase of life science: Schleiden, Schwann, Remak, Virchow.) Thus began cell biology, also known as cytology, which forms one of the intellectual threads that has led to Dolly and beyond. This is the thread that Keith in particular has followed.

How, though, does the single, initial cell – the fertilised egg – multiply to form an entire, multicellular organism, like a sheep or a human being? What happens? What are the underlying mechanisms? Alongside German-based cytology grew German-based embryology. At first this new discipline was perceived as a branch of anatomy but it was established as a speciality in its own right by the German–Estonian Karl Ernst von Baer (1792–1876). In 1827 in *De ovi mammalium et hominis genesi* (*On the Origin of the Mammalian and Human Ovum*) Baer described how all mammals, including human beings, develop from eggs. He showed, too, how the different organs form as the embryo develops – in which order they arise, and from which tissues. He also established comparative embryology – revealing the similarities and differences in the development of different creatures.

Thanks to von Baer and his followers, biologists by the early decades of the 20th century could describe in fine detail how embryos originate and unfold. Each begins with the fusion of female egg (oocyte) with a male sperm – which is the act of conception – to form a single-celled embryo (zygote). The zygote then divides – division is called *cleavage* – and divides again to form a ball of cells, known as a *morula*, which grows and divides further to form a *blastocyst*. The individual cells within a morula or blastocyst are called *blastomeres*. Many mammalian embryos, including those of humans and mice, become embedded or '*implant*' in the wall of their mother's uterus at the blastocyst stage; but those of ruminants, including sheep and cattle, do not implant until the fourth week after conception, when they are several inches in length. Most of this relatively huge structure is *trophectoderm*, however, which as we will see is the tissue that becomes the placenta: the embryo itself forms from only a small part of the structure. As division proceeds further the blastocyst typically becomes hollowed out to form a *blastosphere* or *blastula*. The blastula then folds in upon itself – this is often compared to a tennis ball being pressed in – to form a multilayered but still basically spherical entity now called a *gastrula*.

It is at the gastrula stage that the embryo truly begins to mould itself into the form of the creature it is to become. The whole structure becomes more complicated as the newly formed, discrete cell layers start to slide over each other and fold in upon themselves in a supremely choreographed and self-mobilising exercise in origami. At the same time, different groups of cells in different parts of the embryo can be seen to differentiate, giving rise to daughter cells that are different from themselves and assuming the structure and function of specialist cells in specialist organs – liver, lung, brain, intestine, muscle, and the rest. The transformations from single cell to ball of cells to hollow ball to apparently self-forming creature are intricate, swift and usually unerring – although not without diversion; such that human embryos in their early stages develop but quickly abandon fish-like gill-pouches. Altogether the whole unfolding is stunning. No wonder so many biologists have been drawn to embryology. Thus Ian found his vocation gazing at embryos in Chris Polge's lab while still a student, his thoughts of farming finally evaporating.

But although the development of embryos could broadly be described by the early decades of this century, it could not be explained. How do the cells move? How do they 'know' where they are going? When they get to their final position, how do they 'know' that they ought then to differentiate into lung, or liver, or bone, or whatever? Do they 'know' what they are supposed to become before they set off for their final destination, or do they receive instructions on arrival, and if so what form do the instructions take and where do they come from? Even now these questions are very far from being answered. Science moves rapidly, but big questions take a long time to answer. Again, one prime reason for developing cloning technology is to address these issues. It all simply remains wonderful – and will seem even more wonderful when answers have been provided. Science does not diminish the awe that human beings feel for nature, as its critics have sometimes suggested.

One obvious and key unknown was, and in detail remains, that of differentiation. The finished animal body contains many hundreds of different kinds of cell – liver, muscle, nerve, bone, blood, skin, whatever. Yet each, as Virchow first proposed and now is known beyond doubt, develops by simple division from the initial, single, fertilised egg: the zygote. The zygote looks and behaves nothing like a muscle cell, or a blood cell, or any other kind of cell except itself. Yet it has the potential to give rise to all of those cell types, and many more. It is indeed totipotent. But what are the mechanisms that cause the direct linear descendants of the zygote to change their character so radically? When we know that we can ask more advanced questions, like 'What are the signals that invoke that mechanism?' To make Megan and Morag, and then Dolly, we developed a technique to reprogramme cells that were already differentiated – to re-establish totipotency. We have the technique, and as described in chapter 8, we have some idea how and why the technique works. But we do not understand the details. We would dearly like to. Even when the details are understood, we would not recommend human cloning. We do suggest, though, that *until* the mechanisms of genomic reprogramming are understood a great deal better than they are now, then any contemplation of human cloning must be beyond the pale. More of this later.

To return to our thread: the first significant attempt to explain the mechanism of differentiation was made by yet another German biologist, August Weismann. His explanation seems to be wrong, as it happens, but he inspired others to frame experiments that seem to have supplied the right answer. Thus Weismann can be said to have begun the sequence of work that has produced Dolly.

The beginning of modern times: Weismann to Spemann

August Weismann (1834–1914), professor at the University of Freiburg, is best remembered for his proposal that the *germ cells* (the gametes – eggs and sperm – and the cells that give rise to the eggs and sperm) develop in the embryo quite separately from the *somatic cells* (the body cells). This means that the fate of the body does not, in fact, directly affect the inheritance of the offspring. Thus Longfellow's blacksmith may well have toiled 'from morn till night' and so developed 'brawny arms [as] strong as iron bands' but he would not have passed on his hard-won brawn in his sperm. If his children wanted arms like iron bands they would have to do their own toiling. This is a key observation since many people in the 19th century felt in their bones – following the theories of the great late-18th-century biologist Jean-Baptiste Lamarck (1744–1829) – that creatures do inherit the characteristics that their parents have acquired in life; and Weismann's hypothesis, if correct, would show that there is no mechanism that could enable this to happen. In fact Weismann proposed this idea without good evidence, but the vast weight of observation since suggests that he was absolutely right.

Weismann also considered the mechanism of differentiation but in this, although his reasoning was as always unimpeachable, he turned out to be wrong. He began by pointing out, undeniably enough, that since a zygote does give rise to daughter cells of all the necessary tissue types then it must contain all the hereditary information needed to create all those different types; it is indeed totipotent. But, he went on to suggest, each daughter tissue must retain only a proportion, a sample, of the total hereditary material: only that part of the hereditary material that is appropriate to the particular tissue. Weismann put this idea forward in 1892, and at that time biologists generally tended to suppose that some kind of material, containing hereditary instructions, must be passed on from parents to offspring, but they had no good idea of what that material was. Gregor Mendel had proposed in the 1860s that heritable characters were conveyed by discrete 'factors', but nobody had taken any serious notice and his work did not become known until about 1900. The words 'gene' and 'genetics' are 20th-century inventions. So Weismann, in the early 1890s, must have thought of 'hereditary material' in somewhat general terms; but we would say 'genes'.

Be that as it may, Weismann perceived differentiation as a steady *loss* of hereditary information (genes) within each specialised tissue, as cell division proceeded. He proposed, furthermore, that the loss begins at the very first cleavage, as one cell becomes two. One of the two daughter cells, he supposed, would contain the information for the right side of the body; and the other would contain the hereditary material appropriate to the left; and so on and so on until the highly specialised tissues of nerve, muscle, liver, and the rest each contained just a small proportion of the hereditary whole – only what they needed for their own purposes.

Weismann's proposal was good science since it immediately suggested the means of testing it: take just one of the cells from a two-cell embryo and see if it goes on to give rise to an entire embryo, or only to half of one. So Wilhelm Roux (yet another German embryologist) carried out what seemed to be the appropriate experiment. He decided to work with frog's eggs, which are big, easily available, easily accessible, are able to put up with a great deal of physical insult, and therefore have become a favourite with embryologists. Roux destroyed one cell of a two-cell frog embryo with a red-hot needle. In fact, the remaining cell did produce only half an embryo, so it seemed that Weismann was right again. Roux went on to found a journal, *Entwicklungsmechanik* – 'developmental mechanics'.*

But Roux's experiment, though eminently logical in design, was in practice misconceived, and the answer it gave was erroneous. Yet another German, Hans Driesch, tested Weismann's hypothesis in a different way. He was working with sea-urchin eggs, which

---

* The word 'mechanics', in this context, is provocative. It stands in opposition to the concept of 'vitalism'. Vitalists maintain that life is driven by unique processes that are not explicable purely by the standard laws of physics and chemistry, while 'mechanists' maintain that life simply requires complicated chemistry. This was a very hot issue in the 19th century.

are smaller than frog's eggs, so he could not have repeated Roux's experiment even if he had wanted to: he could not simply have burned away one cell of a two-cell embryo without destroying the other. Instead he shook the two cells apart, leaving both undamaged. And behold: both turned into complete embryos, albeit smaller than normal. He then shook four-cell urchin embryos apart, and produced four complete embryos. But Roux had worked with frogs, and Driesch could not directly address Roux's ideas by his methods since he could not separate the cells of early frog embryos. They are too tightly stuck together.

Driesch's deconstruction of two- and four-cell embryos to make identical sea-urchin twins and quads was probably the first ever successful exercise in human-assisted animal cloning from single cells. Note that his method of producing clones was irreducibly simple: just breaking an embryo apart. Note, too, that Driesch did not set out simply to produce clones of sea-urchins. His motive was to explore the nature of cell differentiation – to discover whether totipotency is lost as cleavage proceeds, and to get some idea of when this happens. But since he found no way to split frog embryos he could not directly refute Roux's conclusions. After all, his results and Roux's might have differed simply because they had worked on different species. The cells of a two- or four-cell sea-urchin embryo might indeed retain totipotency, while those of a frog might lose it. To be sure, sea-urchins are echinoderms, like starfish, and frogs of course are vertebrates, and both belong to a larger animal group called the 'deuterostomes'. This does not prove that sea-urchin and frog embryos should behave in exactly the same way in every respect, though we may feel reasonably confident that observations from sea-urchins are more relevant to frogs than observations from, say, crabs would be. In biology, however, *nothing* should be taken for granted. Yet Driesch thought it likely that cells from young frog embryos would behave like those of young sea-urchin embryos if only they could be separated cleanly – and indeed that his sea-urchin results should apply to all animals. Roux's experiment had misled people, it seems (although this point was made later) because the burnt, dead cell remained stuck to the living one, and inhibited its development. Be that as it may,

Driesch concluded that Weismann was wrong, at least in part. The totipotency of individual cells was not lost irrevocably at the very first cleavage, or even at the second.

Further light was thrown on this crucial issue by one of the great experimental biologists of the early 20th century, Hans Spemann (1869–1941), the son of a bookseller and then, among other things, director of the Kaiser Wilhelm Institute of Biology in Berlin. Spemann was awarded a Nobel Prize in 1935 and was thus the only embryologist to receive the prize before 1986 (although Weismann might reasonably have felt hard done by – after all, Nobel Prizes were first awarded in 1901 and Weismann lived until 1914). Anyway, Spemann was one of those scientists who love the *craft* of science – some of us do and some of us don't – and he was technically wonderful. He worked primarily on amphibians such as newts and salamanders and is best remembered for describing the 'organising centres' of embryos, which direct groups of cells to form particular tissues and organs. He explored the specific phenomenon of differentiation in various ways; one of these was to transplant pieces of tissue from one part of an embryo to another, to see whether the tissue retained the character it began with, or took on the form appropriate to its new surroundings. Thus he found that when *young* tissue is transferred from place to place it changes its nature. This demonstrates that, at least for a time, embryonic tissue does retain some flexibility – an observation which clearly runs counter to Weismann's idea that totipotency is absolutely compromised from the first cleavage.

Most directly relevant in this context, however, were Spemann's experiments with two-cell salamander embryos. He separated the two blastomeres by tightening a noose between them – a noose made from a hair from the head of his baby son. Each of the two blastomeres developed into a whole embryo. Whereas the sea-urchins that Driesch worked with are very different from the frogs that Roux had studied (even if they are both deuterostomes), salamanders are very similar to frogs. Both are amphibians. So what applies to salamanders surely applies to frogs. This work, then, clearly did negate Roux's observations. The blastomeres of two-cell amphibian embryos clearly do retain totipotency.

For all the brilliance of the scientists involved, however, the experiments described so far do not seem to tell us much that any biologist could not have inferred. After all, young embryos do divide in the mammalian womb to produce identical twins, and sometimes even identical quads (or triplets if one of them dies). It was already obvious, then, even if not spelled out, that the totipotency of individual cells is not lost after the first cleavage, or even after the first two. But it is also clear that the blastomeres in a two-cell or a four-cell embryo are not differentiated. So although Weismann might not be correct in all respects – the cells of a two-cell embryo are not already committed to be left-side and right-side – he could still be correct in principle. It could be that genes are lost when differentiation begins, so that differentiated cells do lose their totipotency forever.

Meanwhile, however, a near-contemporary of Spemann's, the German–American Jacques Loeb (1859–1924) at the University of Chicago, was again exploring early development and differentiation in sea-urchins; and in one set of experiments he induced them to undergo parthenogenesis. However, the particular sea-urchins that Loeb studied do not normally develop parthenogenetically. Their eggs need to be fertilised first. To trigger parthenogenesis, then, Loeb had to *simulate* fertilisation. The success of this simulation depends on the general fact that, in nature, the signals that pass between organisms, or between cells within an organism, or between different parts of a cell, are generally minimalist and are always arbitrary. That is, an egg begins to divide and develop when it 'thinks' it has been fertilised, but it does not reach this conclusion by registering in detail the actual penetration and entry of a sperm. It merely registers the disruption of its outer membrane that such entry causes. So, to simulate fertilisation, it is necessary only to mimic this disruption. This can be done, Loeb found, simply by altering the chemical composition of the seawater in which the sea-urchin eggs were kept. All he needed to do, in fact, was to add magnesium chloride – magnesium chloride being a natural component of seawater, although normally present only in very small amounts. Chemical alteration of the seawater in this way alters the electrical properties of the cell membrane, which is disruptive enough to imitate penetration. Note, later, how we control a somewhat different phenomenon – the 'activation' of oocytes – by a comparable technique: altering the concentration of calcium in the surrounding medium.

Anyway, Loeb found that, with such simulation, parthenogenesis proceeded. The unfertilised egg cell began to divide to produce embryos that sometimes developed well. Sometimes, however, adulteration of the seawater disrupted the membrane too much, and it ruptured. When this happened, some of the cytoplasm within would break through the membrane to form a kind of hernia. However – serendipity again! – Loeb was able to turn this peculiar sequence of events to advantage. For sometimes a nucleus from the dividing mass of cells in the main part of the embryo would find its way into this protrusion of cytoplasm. Then, effectively isolated from the rest, it would continue dividing. But – and this really was the surprise – such a nucleus, in this novel, semi-sequestered setting, would sometimes form an entire new embryo.

Two clear lessons emerge from this. The first is that embryonic cells may retain totipotency even after they have undergone several divisions; they certainly do not lose it irrevocably with the first cleavage or even with the first few cleavages. The second lesson is that differentiation, or the lack of it, is heavily influenced at least initially by the cytoplasm. For the nuclei in the main mass of Loeb's embryos were already preparing to form the various kinds of adult tissues. But the cytoplasm in the herniated protrusion apparently retained the properties of egg cytoplasm, and when one of the maturing nuclei wandered into this pristine environment it was induced to behave once more like the nucleus of a zygote. Loeb's was a peculiar experiment which began with an accident, but it led to highly intriguing results.

So then Hans Spemann sought to re-enact the principle that Loeb had demonstrated, this time with his favourite salamander embryos. But he did not simply create crude hernias, as Loeb had done. Instead, he again employed nooses of baby-hair to squeeze salamander zygotes into the form of a dumbbell – with the nucleus in one bell, and unoccupied cytoplasm in the other. He allowed the nuclei to undergo four divisions so there were sixteen cells: the first division producing two cells, then four, then eight, then sixteen. Then he slackened the noose to allow one of those nuclei to travel into the unoccupied cytoplasm, and then he tightened the noose again. This isolated nucleus continued to divide, but, finding itself in cytoplasm that retained the properties of zygote cytoplasm, it again divided to form a complete embryo. Clearly, then, totipotency was retained even after four divisions. Genes were not lost in these early divisions. Clearly, too, whether totipotency was restored to those cells, or the process of differentiation was continued, was determined by the cytoplasm.

There is a conceptual leap between the experiments of Driesch and the early Spemann on the one hand, and those of Loeb and the later Spemann on the other. Driesch and the early Spemann created clones merely by splitting very young embryos, to give identical twins or quads. But Spemann in his later work coaxed nuclei in the cells of older embryos to move into new cytoplasmic environments – nuclei that are the great-granddaughters, or great-great-granddaughters, of the original zygote nucleus. The nuclei are not actually removed from one cell and placed in another: merely shuffled from place to place in the same organism – but from cytoplasm that has matured into cytoplasm that retains the properties of the original egg. Here, then, is the first stirring of nuclear transfer, which is the modern method of choice for cloning, and by which we have now produced Megan and Morag, Dolly, and the rest. Note, too, that although Driesch and Spemann produced clones (even if the clones were not always separated) they did not set out expressly to do so. The cloning was incidental: part of a larger endeavour to study the mechanisms of development, and in particular that of differentiation. Cloning *qua* cloning is almost incidental for us, too. For us, as with most of our predecessors, cloning has been a route to greater ends.

Finally, in 1938, Spemann proposed a 'fantastical experiment' which in truth was a logical extension of the protocol that he and Loeb had already carried out. It would be salutary, he said, to take the nucleus from a differentiated cell – even from an adult cell – and place it in the cytoplasm of an egg whose own nucleus had been removed (that is, an egg that had been *enucleated*). But he did not carry out this experiment because, he said, he could 'see no way' to introduce 'an isolated nucleus into the protoplasm of an egg devoid of a nucleus'.

Note that Spemann's 'fantastical experiment' really was fantastical. Was it actually possible to rip the nucleus from one cell and put it into another? On the face of things this seems too traumatic even to be contemplated: like ripping the brain from a body and thrusting it into another. Would the brain and the recipient body establish a harmonious relationship? Surely not. Would any of the participants survive – the donor, the recipient, the organ itself? Surely not, again. Nuclear transfer is effected by what is called 'microsurgery' but from the viewpoint of the cell a more major insult can hardly be imagined.

Yet such transfer was attempted – and sometimes, surprisingly often in fact, the nuclei do survive, and the recipient and the transferred nucleus do establish a perfectly harmonious relationship, as if they were naturally intended for each other. Success has been greatly increased by a series of technical refinements – not the least of which is an agent called cytochalasin B, which relaxes the cytoskeleton, and reduces the damage done by the transfer – but the wonder is that it succeeds at all. This is yet another of nature's serendipities – comparable with Buchner's discovery, late in the 19th century, that enzymes can often function perfectly well outside the cocoon of the living cell. There seems to be no *a priori* reason why nuclear transfer *should* work. If it had proved as traumatic as the transfer of a brain then cloning of the kind that has enabled us to produce Dolly and the rest would surely be a non-starter. But this particular exercise did work, and the research has gone on. We should not forget, though, that Spemann was right. That this work succeeds at all *is* 'fantastical'.

But we are running ahead. Others did take up Spemann's challenge, and truly began the age of nuclear transfer that he and Loeb had presaged. Megan and Morag, Taffy and Tweed, Dolly and all their successors are the latest exemplars. But there were crucial intermediate stages, beginning with Robert Briggs and Thomas King in America in the 1950s, and continuing with

John Gurdon in England in the 1960s. As a student Keith was particularly inspired by Gurdon's work.

The age of nuclear transfer: Briggs, King and Gurdon

In the 1950s Robert Briggs worked for the Institute of Cancer Research in Philadelphia. In cancers, the normal genetic programme is in disarray. Cells that ought to become differentiated proliferate instead, somewhat in the manner of rampant embryonic cells, but without discipline; indeed, one theory says that cancers *are* rampant embryonic cells. Thus the ill-discipline of cancer and the discipline that turns a zygote into an animal with a multitude of tissue types, all perfectly positioned, seem to reflect opposite sides of the same coin. The abnormality of cancer should throw light on the normality of embryo development, and vice versa. As we will see, Keith made a very similar observation when he studied the biology of cancer during his years at the Marie Curie Memorial Institute.

Weismann's hypothesis – that cells lose genes as they differentiate – was still very much on the agenda by the late 1940s and early 50s. Loeb's and Spemann's experiments certainly showed that totipotency is not lost with the very first divisions, as Weismann had suggested might be the case. But then, Weismann was mainly concerned to provide an explanation of differentiation – and there is no visible sign of differentiation after the very first divisions in an embryo. Indeed, differentiation cannot be clearly *observed* – at least under the light microscope – until the embryo is past the blastula stage, and has become a gastrula. Perhaps the differentiating cells begin to lose genes at that point. On the other hand, there is an obvious alternative hypothesis: not that genes are lost as differentiation proceeds, but that they are simply shut down. Strange to relate, Weismann himself thought of this idea – but then rejected it. The history of science is full of such oddities, and near-misses. In any case, this is the specific hypothesis that Briggs set out to test.

At the time that Briggs began his experiments he had no detailed idea of the mechanism by which genes might be shut down. Indeed, it had only just been made clear (by Oswald Avery and his colleagues in America) that genes are, in fact, made of DNA and not, say, of protein; and Francis Crick and James Watson had yet to produce their three-dimensional model of DNA (which they did in 1953). Genes, in the late 1940s, were still seen simply as abstract entities, albeit entities that influenced body form in precise and quantifiable ways. Still, it was possible in a general way to infer how they must be turned on and off during development, assuming that this was the case.

The first step in science is to frame a good question, and the next step is to test it. As the great British immunologist Sir Peter Medawar observed, adapting an epigram of Bismarck, 'Science is the art of the soluble.' So – how to test the idea that genes are simply turned off during development, and not irrevocably lost? Briggs's colleague, Jack Schultz, suggested a solution – a direct echo of Spemann's proposed 'fantastical experiment': transfer the nuclei from mature somatic cells of frogs into enucleated frog eggs. If the nuclei from these differentiated, or quasi-differentiated, cells redeveloped into complete embryos in the pristine cytoplasmic environment of an egg, then they could not have lost any of their genes, and differentiation clearly involved a mere switching off. Unfortunately, of course, if they failed to develop this would not prove that genes had indeed been lost; it might simply mean that the nuclear transfer had been too damaging. But then, no experiment will ever tell you everything you want to know.

Frog eggs, as John Gurdon was later to observe, are 'remarkably resistant to microsurgery', and this is another piece of serendipity, for if they were not (and there is no known law that says they are bound to be) then this whole field of research would have come to a stop, or at least been obliged to take a wide diversion. Anyway, Briggs secured the services of the extremely dextrous Thomas J. King, from New York University, and in 1950 they began work together on the common American frog *Rana pipiens*. They first had to remove the nuclei from the frog eggs; and this is possible because (luckily) the nucleus is well to the side of the egg, in fact just under the surface, and so can be seen through the microscope and teased out with a needle (or destroyed by ultraviolet light).

Then, with a pipette, they sucked out cells from frog blastulae that had 8000 to 16 000 cells – not yet obviously differentiated, but well past the first few divisions. The pipette was wide enough to accommodate the cell nuclei easily, but too narrow to accommodate the whole cell. Briggs and King anticipated that if they left the cell membrane intact, then, after it was transferred to the enucleated egg, signals from the egg cytoplasm would not get through to the nucleus. On the other hand, if they stripped the cytoplasm entirely from the transferred cell, then the nucleus would be damaged. This was clearly one of the more ticklish stages of the whole operation: to remove the obstructive cell membrane without wounding the cell beyond recovery. A simpler technique has evolved since then, and when we made Megan and Morag and Dolly we did not strip away the membrane of the donor cell in order to fuse it with the receiving cytoplast, but simply disrupted the membranes of the two cells by electric pulses (a technique that is conceptually comparable, though superficially very different, from the chemical disruption by which Jacques Loeb persuaded sea-urchin eggs that they had been penetrated by sperm). Be that as it may, John Gurdon was later to observe that, until Briggs and King came along, 'the only kind of nucleus that could be made to penetrate an egg was the nucleus of a sperm cell'. Embryos formed by transferring a new nucleus into the enucleated cytoplasm of an egg or zygote are sometimes called 'reconstructed embryos'. In our 1997 paper on Dolly we coined the term *'couplet'*.

Briggs and King had their first success in November 1951, when one of their reconstructed frog embryos survived and began to develop – though only to be crushed by the over-eager forceps of a visiting scientist. But in the end, out of 197 reconstructed embryos, 104 began development, 35 became embryos, and 27 grew into tadpoles.

Briggs and King had thus shown that the cell nuclei of blastulae that were well past the two- or four-cell stage retained totipotency, which could be made evident if they were placed in an appropriate cytoplasmic environment. They had also produced the world's first clones by the method of nuclear transfer – which is conceptually quite different from, and more advanced than, the mere splitting of embryos. We cloned Megan and Morag, Taffy and Tweed, Dolly and Polly and their like by nuclear transfer and so they are the direct conceptual descendants of Briggs's and King's tadpoles. Briggs's and King's humble, short-lived protégés truly began – if any single instance can be said to have done so – the modern 'clone age'. Note yet again, however, that it was not their prime intention to produce clones. Cloning was simply the means by which to study differentiation. Cloning research has sometimes been described as if it were simply an exercise in commerce – at attempt to multiply particularly valuable livestock. But it has been driven at least as much by the desire to address fundamental questions of biology, which the technology is uniquely equipped to do. Even in this harsh, modern commercial world, this remains a vital subtext of our own research.

Others followed in the footsteps of Briggs and King and by the time John Gurdon began his nuclear transfer work at Oxford in the late 1950s and early 1960s he had plenty of good information to draw upon. It was clear that, in the early stages of embryo development, the cytoplasm is in the driving seat. The cytoplasm induces the first cell divisions in the embryo, and determines what course the daughter cells will follow. Egg cytoplasm could redefine the fate of blastocyst nuclei that were transferred into them. But, Gurdon observed, embryos reconstructed by nuclear transfer rarely developed beyond the gastrula stage – the stage at which differentiation begins in earnest. At this stage, so biologists inferred, the genes – the DNA – in the transferred nuclei took over the development. Failure at this stage, then, implied that the DNA had been damaged, or that it could not function as it should within the new cytoplasmic environment, or that dialogue between transferred nucleus and host cytoplasm was not all that it should be.

So, as Gurdon pointed out, Weismann's notion – that differentiation involves loss of genes – was still not dead and buried. It remained possible that cells began to lose genes at the time when differentiation could clearly be observed – which means at the gastrula stage. To show that Weismann was wrong, then (meaning that some other explanation had to be right) Gurdon wanted to show that nuclei from cells that were clearly differentiated – post-gastrula cells – retained totipotency, and could produce whole animals when transferred into egg cytoplasm.

Gurdon worked with the African clawed frog, *Xenopus laevis*, the much favoured laboratory animal that Keith also employed for his doctoral thesis. Gurdon transferred cells from the intestinal lining of tadpoles that had already begun to feed. But this immediately raises two theoretical problems. First, the fact that the tadpoles were feeding showed that their guts were already specialised, meaning that the cells they contained were already differentiated. But it is at least theoretically possible that *some* of the cells within the gut – or indeed in any tissue – retain the properties of embryo cells: that they are, in fact, simply cells that have multiplied without differentiating at all, but turn up here and there in mature tissues, or, more specifically, that they are stem cells. As we will see, this remains theoretically possible even in the case of Dolly. However, at least some of the cells that Gurdon transferred were clearly differentiated, since they had the kind of absorptive surface – a 'brush border' – that is typical of mature gut.

Secondly, an embryo may indeed develop from a reconstructed embryo – but did the embryo really develop from the nucleus of the transferred cell? It is obviously possible to remove or destroy the nuclear material of a frog egg, but it is not exactly easy, and it is at least possible that the nucleus might be left behind. Then, the embryo might simply develop from the original nucleus, and there would be nothing unusual about that. Gurdon, however, transferred cells from tadpoles that carried a genetic mutation that led to albinism – lack of pigment – which made it easy to see that any embryos that developed did indeed derive from the transferred cells. Similarly, we made Dolly by transferring a Finn-Dorset nucleus into the enucleated egg of a Scottish Blackface ewe. The two kinds of sheep look quite different. No one can doubt which participant supplied the genes. Nowadays, too (as described in chapter 10), the genetic provenance of any creature can be determined well beyond reasonable doubt by 'genetic fingerprinting'.

Anyway, Gurdon showed that at least some of the nuclei that did derive from tadpole intestinal cells – including some that were differentiated – could produce whole embryos when transferred into enucleated *Xenopus* egg cytoplasm. Furthermore, some of those embryos went on to become tadpoles, and some of these even turned into frogs. Indeed, he said, 'Both male and female adult frogs, fertile and normal in every respect, have been obtained from transplanted intestine nuclei.' In fact only about 1.5 per cent of the transferred intestinal cells survived to be adult frogs; but all of these carried the genetic marker which showed they had indeed derived from the transferred nuclei. Gurdon later showed, however, that a much greater proportion of transferred cells were able to sustain development at least to the stage where tadpoles start to twitch – a stage at which they clearly possess the highly differentiated tissue of nerve and muscle. In general, he concluded, transferred cells that are more specialised are less likely to support development beyond the early stages than are less specialised cells. This remained one of the central ideas in cloning biology right up until the mid 1990s. In essence it is surely right – although the birth of Megan and Morag in 1995 has produced a crucial change of emphasis. All in all, Gurdon produced fully functional adult frogs from specialist tadpole cells but he never managed to produce adult frogs from cells transferred from adult cells. But that is what we achieved with Dolly.

Gurdon's experiments still do not quite disprove Weismann's hypothesis, for it is possible that at least some of the most specialist cells do indeed lose genes as differentiation proceeds. This would explain why they are less able to support development in the embryos that are derived from them. Yet his work does strongly suggest that differentiation is actually effected in some other way – the most obvious hypothesis being that genes are simply switched off as cells specialise. In fact, the switching on and switching off of genes is now to some extent understood and this surely is the principal mechanism. It does seem that in some cells differentiation may involve the physical rearrangement of genes – as in the stem cells of the immune system – or indeed their physical loss. The final stage of differentiation in the red blood cells of mammals involves the loss of the entire nucleus – casting out *all* the nuclear DNA. In general, though, differentiation does *not* involve the physical loss of genes.

So the pioneer observations and theories of the 19th-century cell biologists and embryologists represent the first stage on the path to Dolly; the more intrusive investigations from Roux to Spemann represent stage two; and the nuclear transfers of Briggs and King, and then of Gurdon, represent stage three. Our work – first creating embryos from cultured cells and then deliberately re-establishing totipotency – can be said to represent stages four and five. John Gurdon's demonstration that specialised tadpole eggs can give rise to embryos that will grow into fertile and healthy adult frogs shows that at least *some* specialised cells do retain totipotency (or at least to give rise to totipotent daughters), and this ability is made manifest if they are placed in a suitable cytoplasmic environment.

But all the work so far has one obvious limitation. Frogs and salamanders are amphibians. Human beings are mammals and so are most of the other species that concern us most, including the farmyard creatures: cattle, sheep and pigs. Keith emphasises that, at the level of the cell, organisms tend to be much the same, and this has proved a vital insight. In similar vein, it could be that all the rules of development that apply to amphibians also apply precisely to mammals. But nature cannot be second guessed, and we cannot assume such similarity. Besides, if we want to translate science into serious technology, and apply what we know to agriculture and to human medicine, then we *have* to work with mammals.

But the qualities of amphibians that make them easy to work with emphatically do not apply to mammals. Mammal eggs are tiny: at least, they are huge compared with most other mammalian cells but minute against those of frogs, like ping-pong balls against pumpkins. Amphibian eggs are 1200–1500 microns in diameter (a micron is a thousandth of a millimetre – so frog's eggs are big enough to measure with a school ruler). A mammalian egg is only around 100 microns across – that is, a speck, just visible to the naked eye (while most body cells are about 15–20 microns in diameter). Worse: the eggs of mammals (apart from those of duck-billed platypuses and echidnas) never see the light of day. Fertilisation is internal. So is development: first in the oviduct and then in the uterus. It is precisely because amphibians externalise their reproduction so flagrantly and vulnerably that they are still living in ponds. Mammals are as successful as they are largely because they keep their embryos so snugly cocooned, deep and moist within their own bodies. Human babies at birth are fairly helpless but they are already well formed. Lambs and calves at birth are almost ready to take their place in the herd; young calves, to be sure, spend a lot of time resting but they can stand to suck milk from day one. In short, by the time a baby mammal enters the outside world, and so presents itself for observation, most of development has already taken place.

So, to study the reproduction and early development of mammals, biologists needed to develop a whole new range of techniques that would bring these most secret events into the open, encourage them to happen in the petri dish rather than the ovary or womb, and/or find better ways of exploring the animals' reproduction internally. The peculiarities of mammalian reproduction, and the first essays in mammalian cloning, are the subject of the next chapter.

CHAPTER 5

The Facts of Life Revisited

Ian Wilmut and Keith Campbell:

It is impossible to understand the making of Dolly, and Megan and Morag, what we did and why we did it, without discussing details of mammalian reproduction. We don't just put donor nuclei into generalised 'eggs', for example. We put them very specifically into 'MII oocytes' (pronounced 'em-two oh-oh-sites'). If you know what an MII oocyte is, how it comes about, and how it differs from, say, an 'ovum' or any of the other entities to which the broad term 'egg' is applied, then please skip this chapter. Take warning though: for although children these days are taught the alleged facts of life in primary school, and all adults claim to be experts, most people when quizzed get most of the details wrong. Biologists may know the technical vocabulary – meiosis, spindle, zona pellucida and the rest (these terms are explained later) – yet they too are hazy on vital points unless they are reproductive specialists. Some textbooks are certainly wrong, either through deliberate and carefully planned simplification or – banish the thought! – because their authors are confused as well. So this chapter is probably inescapable. But, as William Makepeace Thackeray used to tell his readers – 'Don't worry! It gets better!'

What people believe and what is the case

The basic idea of sexual reproduction is as related in chapter 3. The standard body cells of most animals are diploid; that is, they contain two complete sets of chromosomes, which means two complete sets of genes, one of which is inherited from the mother, and one from the father. The two sets are said to be *homologous*: each chromosome in each set has its 'homologue' in the other. Most animals reproduce by sexual means (although a few practise cloning as well and some have abandoned sex altogether) and to do this they produce gametes (specialised sex cells) that are haploid; that is, they contain only one set of chromosomes. In sexual reproduction a female gamete (an egg) joins with a male gamete (a sperm) in the act of fertilisation, to produce a diploid cell. This diploid cell is, in effect, a new individual. It is the one-celled embryo known as the zygote.

Such a narrative certainly conveys the basic idea. A diploid body produces specialist haploid cells which join with other haploid cells to produce diploid cells again. But such a simple account implies a great deal which just is not the case. It implies, for example – the point seems obvious – that at the time of fertilisation the female gamete is haploid. Textbooks commonly speak of the haploid 'ovum'. But in fact, at the time of fertilisation, the female gamete is still diploid; it is still, indeed, at the stage known as the 'oocyte'. The oocyte completes the progression to haploidy *only* when triggered to do so by the sperm. The sperm penetrates the zona pellucida (the thick, transparent layer surrounding the oocyte, which is described in detail later), then makes contact with the outer membrane of the oocyte, and by making contact *activates* the oocyte – which means that it prompts the second meiosis to run to completion, with expulsion of a ball of surplus chromosomes contained in a membrane, known as the 'second polar body'. This is an odd fact, and it may seem a small one, but in the context of cloning by the methods applied at Roslin it is highly significant.

The standard account, too, implies that after fertilisation the nucleus (now called a 'pronucleus') of the sperm joins with that of the egg to produce a zygote that contains one diploid nucleus: a glorious coming together of hereditary material, like political delegates at a preconference ball. It's a fine vision. It is what most people seem to understand to be the case. But such congress never actually takes place. In practice, after fertilisation the female and male pronucleus – both haploid – sit decorously side by side in the cytoplasm of what is now the zygote, and replicate their chromosomes. Not until the zygote is preparing to divide to become a two-cell embryo do the two sets of chromosomes from the two pronuclei begin to act in coordinated fashion. But the two sets of chromosomes from the two parent gametes do not come together within one nucleus until the embryo is at the two-cell stage. Again, this may seem odd; yet, when you think about it, the mechanism by which the maternal and paternal chromosomes are brought together is efficient. There is no point in the two sets of chromosomes from the two parents mingling together before they have duplicated, and before the zygote is ready to divide; so, in practice, they don't.

Then there is the matter of growth. The passage of an animal from one-celled zygote to myriad-celled, free-living individual clearly involves two processes: cell division, and overall growth. It is natural to assume that both processes begin together – as soon as the zygote first divides. But, again, this is not so. The first cell divisions in an animal embryo take place without any growth at all. Depending on the species, an embryo with several thousand cells may be no larger than the original zygote (and indeed may be smaller, since some materials will have been lost through excretion). Of course, the nuclei have multiplied, so the total mass of nuclear material is much greater; but the raw materials for nuclear increase are supplied by the cytoplasm of the zygote – which means by the cytoplasm of the original oocyte, since the sperm provides almost none.

The cytoplasm itself divides into smaller and smaller portions, as cell division proceeds without total growth. The resulting cells become smaller and smaller. But then, typical body cells are a mere 20 microns or so in diameter while oocytes are large – 100 microns across in a human or a sea-urchin, a millimetre or more (1000 microns) in a frog, several centimetres in a bird. A fivefold difference in diameter between a mammalian body cell and mammalian oocyte implies at least a 100-fold difference in volume. In general, cell division proceeds without total growth until the cells are reduced to the typical body cell size of the species in question. After that point the cells cannot divide further unless they also grow between divisions. So the embryo begins with a process of divide, divide, divide; and only later enters the pattern that will persist throughout life – grow–divide, grow–divide, grow–divide.

But enough has been said to show that the details of reproduction are not quite as most people, even most biologists, understand them to be. The only way truly to appreciate our work, and all that came before and might come after, is to go back to the basics and see what really does happen.

The basics: how cells divide

The division of animal, plant, and fungal cells is of necessity intricate and in practice is a miracle of organisation, as precise as a naval regatta – and, in the behaviour of the chromosomes, resembling one. The genetic material has to divide *exactly*. August Weismann thought that as cells divide and differentiate so they lose hereditary material (or genes, as we would call them now) but now it seems virtually certain that this is not the case, except in some exceptional circumstances. After division, each daughter cell must finish up with precise copies of the parent cell's genes – implying, of course, that each daughter cell contains precisely the same genes as its twin. Genes are carried on chromosomes and the act of apportioning precise assignments of genes between the two daughter cells can actually be watched under the microscope, as the chromosomes can be seen dividing. Logic tells us, however, that the chromosomes could not divide to give precise copies of what was there before unless the material within them had first been doubled; and indeed there is a crucial phase in the life of the cell (known as the S, or synthesis, phase) when the DNA in the chromosomes replicates. During the replication phase, however, the DNA is spread out throughout the nucleus, and is not visible through the light microscope. Unfortunately, again, old-fashioned accounts of cell division tend to focus on what is visible (the division of the chromosomes) and not on the essential replication of DNA that occurs out of sight in between divisions. But note that the replication of cells requires two processes: the replication of DNA, which cannot be observed through a light microscope; and the subsequent division of the chromosomes, which can.

Of course, the cytoplasm has to divide as well – not just the chromosomes. But the division of the cytoplasm does not need to be quite so precise. So long as the basic items of apparatus are doubled up more or less – mitochondria, Golgi bodies, and so on and so on – so that there is enough in each daughter to get things started, that seems to be good enough. In growing body tissues, where each cell grows between divisions, the cytoplasmic material needs to be (roughly) doubled during the growth phase, and this doubling needs to be coordinated (roughly) with the precise duplication of the DNA. But in a young embryo there is no need to coordinate the growth of the cytoplasm with the replication of the DNA since the cytoplasm does not actually grow in a very young embryo.

In practice, animals must be able to operate two forms of cell division. The standard form simply multiplies body cells, which means that diploid cells simply generate more diploid cells, and this is known as *mitosis*. But there is also a special form of cell division designed to produce gametes, which enables the diploid parent cells to generate haploid daughters. This is known as *meiosis*. The two terms sound very similar but they have quite different roots. The word 'mitosis' derives from the Greek *'mitos'*, meaning 'thread', where 'thread' alludes to the visible chromosomes themselves; while the Greek 'meiosis' means 'diminution', and refers to the transition from diploidy to haploidy.

Overall, cells are either dividing (by mitosis or meiosis) or they are between divisions; and when they are between divisions, they are said to be in *interphase*. During interphase the DNA does most of its day-to-day work – making RNA that will help to make proteins – and also replicates. In principle the duplication of DNA is straightforward, yet in execution it is miraculous, although it happens millions of times each second, in each of us. To be sure if we are adults then we will have stopped growing and the cells in some of our most specialised tissues, such as nerves and muscles, have stopped dividing. But the cells in busy and exposed tissues like the liver, gut, and skin constantly renew themselves throughout life (the liver normally does so slowly, unless damaged; but the replacement of gut lining is frenetic).

The two phases together – mitosis (or meiosis) and interphase – are called the *cell cycle*. As we will see in later chapters, there is a great deal more to interphase than is outlined here; and so there is a great deal more to the cell cycle than this present account conveys. Indeed the cell cycle is so complicated in detail that it needs a whole chapter to itself (chapter 8); and such a lengthy treatment is justified because, as it turns out, understanding the cell cycle is the key to cloning by nuclear transfer.

Before we get to this, however, we should look in more detail at the particular processes of mitosis and meiosis. Again, the details are pertinent.

Standard, workaday cell division: mitosis

The genes of all creatures are made of DNA and in eukaryotic creatures like us, and plants and fungi, the DNA is neatly packaged in a series of huge 'macromolecules' known as chromosomes. Each species has a characteristic haploid number of chromosomes, known as n; and this haploid number is of course doubled in diploid cells (so the diploid number is $2n$). In human beings the haploid number n is twenty-three and so each diploid cell contains forty-six chromosomes; chimpanzees have two sets of twenty-four, which makes forty-eight chromosomes; and sheep typically have two sets of twenty-seven chromosomes, which is fifty-four. The complete apportionment of genetic material is called the 'genome'; and with a few exceptions (as in red blood cells) each and every body cell carries a complete set of chromosomes, which means it contains a complete genome.

The particular number of chromosomes in a species is not

related to intelligence or anything else, as far as can be seen, but is just the way things have turned out. However, in general, difference in chromosome number between different species creates a reproductive barrier (usually one of several such barriers). That is, closely related species with different chromosome numbers may in some instances mate successfully to produce offspring; but when those offspring come to reproduce in their turn they usually find they are unable to produce gametes successfully.

Each chromosome contains only one macromolecule of DNA, which is accompanied and supported physically by various chromosomal proteins of which the chief, providing a solid skeleton for the chromosome, are called 'histones'. The DNA and the supporting proteins together are collectively called 'chromatin'. Each DNA macromolecule consists of two helices, neatly intertwined: the famous 'double helix'. During interphase, the macromolecules are spread throughout the whole nucleus; only when they are spread out can they easily fulfil their day-to-day function, which is making proteins, and also replicate themselves ready for subsequent division. When they are spread out they are not visible through the light microscope, but they can be seen through the electron microscope. Their position, when spread out, looks somewhat random but it clearly is not: the position of each piece of DNA – each gene – in three-dimensional space clearly influences its expression. Be that as it may, the macromolecules cannot be seen under the light microscope until they start to coil even more fiercely upon themselves, which they do at the beginning of mitosis (or meiosis) to facilitate the act of division. In this bunched-up state they are said to be *condensed*. It is in their condensed state that they show up as discrete bodies when stained with coloured dyes, and it is then that they are properly called 'chromosomes' – which, of course, means 'coloured bodies'. Remember that the 'A' in 'DNA' stands for 'acid', and the DNA takes up acid-binding dyes. Entities in biology are often named after what the biologist can actually see, rather than what they do. In olden times, at least, things were generally seen before their function was known (although this is not true of the gene itself!).

Now for another point that is often missed – or at least, underemphasised. Drawings or photographs of chromosomes typically show them as *doubled* structures; in fact they look like various forms of crosses, with four branches extending from a single point (which in fact is the *'centromere'*). Normally this passes without comment. But why are the chromosomes cross-like? Why, when each chromosome consists of a *single* DNA macromolecule?

The answer is, of course, that chromosomes are commonly drawn, or photographed, just as they are about to enter mitosis. They are drawn or photographed at this time because this is when they are condensed and hence can be seen under the light microscope. But at the point where they enter mitosis the macromolecules have already doubled – albeit out of sight, during the cryptic S phase of interphase. Hence when they become visible, it is as doubled structures. So, really, the doubled cross-like structure represents two complete copies of the chromosome. To be sure, so long as the two copies remain joined together at the centromere, then each one is called a *'chromatid'*. 'Chromatid' sounds like a diminutive. But functionally, each chromatid is a whole chromosome. Thus, after the S phase, and before the mitosis phase, each body cell contains *four* functional sets of chromosomes. A cell containing four sets of chromosomes is said to be *tetraploid*. Hence although animals are in general diploid most of their body cells spend a significant proportion of their time in a tetraploid state.

Again, most biologists don't seem to have considered this; they just don't think of diploid cells being tetraploid, even though they know the mechanisms of cell division (in broad terms) and common sense says it must be so. But again it is important to the technology of cloning by nuclear transfer that diploid cells are tetraploid for some of the time.

The process of division itself is like a piece of choreography; and, as with all choreography, the different movements are formally named. Stage one is *prophase*. In prophase, interesting events can be seen unfolding both inside the nucleus and in the cytoplasm immediately outside, involving the cytoskeleton. Inside the nucleus, the chromosomes, still encircled by the nuclear membrane, begin to condense and so become visible – as doubled structures – under the light microscope. Outside the nucleus, parts of the cytoplasmic skeleton form themselves into two conical structures, like the frames of two wigwams. These two wigwams lie with their wide ends together. The whole structure is then called the *spindle* – the two pointed ends of which, facing away from each other, are called *poles*. The struts of the wigwam frames are made of protein fibres that are able to contract; by contracting, they will later pull the two sets of chromosomes apart. Prophase, then, is the build-up to mitosis – or, at least, the visible part of the build-up. The chromosomes inside the nucleus get themselves into a coiled-up, condensed state in which they can safely be hauled about; and outside the nucleus the spindle forms, which does the hauling. Of course, the proteins that form the contractile threads of the spindle must initially have been brought into being by the DNA in the nucleus; but – and here is another important principle – now that the threads are formed, they manipulate the chromosomes. The cytoplasmic apparatus thus effects chromosome division. This is a visible example of the dialogue between nucleus and cytoplasm; and in this instance the cytoplasm takes the lead.

Prophase moves smoothly into *prometaphase*. In this stage the nuclear membrane breaks down and disappears from view. The nuclear membrane is also known as the nuclear envelope and '*nuclear envelope breakdown*' has acquired the acronym '*NEBD*'. As again will become apparent, the phenomenon of NEBD and the chemical signals that cause it to happen are vital to the technology of cloning by nuclear transfer. With no nuclear membrane, there is in effect no nucleus: the condensed chromosomes appear as free, doubled structures floating in the cytoplasm. The nuclear membrane will not be re-formed until chromosome division is complete. The re-formation of the nuclear membrane is the penultimate stage of cell division: the last of all being the division of the cytoplasm.

Prometaphase is quickly followed by *metaphase*. With the nuclear membrane out of the way, the chromosomes that were inside the nucleus and the spindle that formed outside it can mingle together. In practice, the centromeres that hold each of the doubled chromosomes together join to the protein threads that form the spindle. They do this very neatly: they all line up along the widest part of the spindle, right in the middle, where the two wigwam halves meet. The closest thing you will ever see to a straight line in the living body occurs at this stage of mitosis, where the chromosomes are lined up line abreast across the central plane of the spindle.

After metaphase comes the most dramatic sequence of all: *anaphase*. In anaphase the centromeres are severed and the two chromatids in each pair are yanked apart from each other This yanking is both a pull and a push: the protein threads that form the spindle contract, and bodily tug the chromatids towards the poles; meanwhile, more protein structures form between the separating sets of chromatids and shove them apart. The net effect of this push-me-pull-you is that one complete set of chromatids finishes up at each pole, each a perfect replica of the other. At this stage, when the chromatids have been separated from each other, they can again be regarded as complete chromosomes. But at this point, immediately after cell division, they are single structures (although each consists of a double helix of DNA). Anaphase begins suddenly and lasts for only a few minutes.

Now that we have two complete sets of chromosomes, it is just a question of restoring order. This restorative phase is called *telophase*, from the Greek '*telos*', meaning end. The nuclear envelope starts to reappear, so now we have two complete daughter nuclei, each with a complete complement of chromosomes exactly like that of the parent cell. The chromosomes now start to decondense and, in their spread-out state, to begin making proteins and replicating themselves, which is the serious business of interphase.

With mitosis over, the cytoplasm can separate to give two complete cells. The cytoplasmic material does not divide with anything like the precision of the nuclei, but this does not matter. So long as each new daughter cell has a reasonable complement of essential parts – mitochondria, ribosomes, and so on – all will be well. Quantities can be adjusted later. The division of the cytoplasm itself is the cleavage process. Basically, the membrane around the middle of the cell, between the two new daughter nuclei, is drawn in until the two sides meet. So now we have two daughter cells where before there was one. Each has the same complement of genes as the other. The cells' cytoplasm may differ somewhat in the quantity of materials they contain but qualitatively the two cells are the same. The two cells together form a clone – or, we may say (such is the flexibility of the term 'clone'), each is a clone of the other.

This, then, is mitosis: the standard form of chromosome division in most of the body cells of most animals, which is designed to produce two diploid daughter cells from a diploid parent cell. Meiosis, by contrast, is designed to produce haploid daughter cells from diploid daughter cells, which means it is intended to produce gametes (eggs or sperm). It is worth taking a brief bird's eye view of gamete production, just to put meiosis into perspective.

The production of gametes

In all vertebrate embryos a clear division arises early on between the somatic cells, which form the body of the individual, and the *primordial germ cells*, which will give rise to the gametes. Exactly what the signals are that tell certain of the cells to become primordial germ cells is not known; but the ones that are so chosen migrate into the developing gonads (testes or ovaries). At this stage of course they are still diploid, and they proliferate within the gonads, by mitosis, to form a *germinal layer*. The diploid germ cells that will give rise to spermatozoa are called *spermatocytes*, and the diploid mother cells of eggs are called *oocytes*.

Later, however, these germ cells undergo meiosis, to produce gametes. In principle and in many details meiosis is similar to mitosis. But since it is designed to halve the number of chromosomes – a diploid cell gives rise to haploid daughters – it is more complicated. Furthermore, although the chromosomes in spermatocytes and oocytes behave in the same way, egg production as a whole has several peculiar and counterintuitive features – which again most people, including most biologists, seem to get wrong. We will first look at meiosis in general (how the chromosomes behave in gamete production) and then look at the special features of eggs.

Meiosis in general

Biologists first became aware that there is such a thing as meiosis in 1883, when it was found that the fertilised egg of a particular worm contains four chromosomes, while a sperm of the same worm has only two. This observation also reinforced the notion that chromosomes are involved with heredity. At that time, unfortunately, Gregor Mendel's work on what he called hereditary 'factors' had been temporarily forgotten, so no one saw immediately that his 'factors', later known as genes, are carried on the chromosomes. Once see that, and the glorious underlying simplicity of biology starts to fall into place. The details may be complicated, however: so much so, that the ins and outs of meiosis were not worked out until the 1930s – 50 years after meiosis itself was discovered.

Thus you might suppose that mitosis could easily be transformed into meiosis just by cutting out, say, the initial duplication of the DNA. After all, there are two complete sets of chromosomes in a germ cell, and reduction division of a kind would be achieved if the two sets simply lined themselves up along a spindle in a form of metaphase, but without duplicating first, and if one of each pair travelled to opposite poles of the spindle.

But this is not how meiosis works at all. In practice, the DNA does first replicate in an S phase, just as in mitosis, to produce two complete sets of doubled chromosomes. But then there follows not one but *two* divisions, known as meiosis division I and meiosis division II or, more simply, meiosis I and meiosis II. Each stage of meiosis – meiosis I and meiosis II – is further subdivided into prophase, metaphase, anaphase and telophase, just as in mitosis; so in total we have prophase I, metaphase I, anaphase I, and telophase I, followed by prophase II, metaphase II, anaphase II, and telophase II. The net result is that each parent germ cell produces not two but *four* daughter cells (although, as we will see, in the female all but one of these are thrown away).

This may seem like gratuitous complexity but in fact the process operates beautifully to mix the genetic material from each pair of homologous chromosomes, to produce a novel arrangement of chromosomes that is unique to each gamete. No two individuals produced by sexual reproduction are exactly alike. Clones, of course, forgo some of this uniqueness.

Meiosis does begin like mitosis. There is an S phase in the primordial spermatocyte and oocyte in which each of the DNA molecules doubles, so that when the chromosomes first become visible during prophase they are already doubled. But now comes an elaboration that is unique to meiosis. Each chromosome pairs up with its homologue; each paternally derived chromosome pairs with its corresponding maternally derived chromosome. Both the chromosomes in each matching pair are already doubled; and the two doubled structures lie very close to one another, touching along their length. So now, in prophase I of meiosis, the chromosomes appear as *quadrupled* structures: four complete chromatids lying side by side. In fact, although the structures are quadrupled, they are at this stage called *bivalents*, since they each contain two complete (doubled) chromosomes.

It is at this stage, in prophase I of meiosis, with four chromatids lying closely side by side, two of paternal origin and two of maternal origin, that we truly see the essence of sex. For the whole point of sex is to mix the genetic material between two individuals. In fact, during all of the life of the cell until this point, each chromosome has remained physically separate from its homologue. Certainly the two interact in various ways; but they remain apart. When they form bivalents, however, each chromosome is in close physical contact with its homologue. And at this point we see sex in action; for each chromatid swaps genetic material with the homologous chromatid lying alongside. This swapping is called '*crossing over*'. In humans there is an average of two to three crossings over between each chromatid pair. The chromatids that emerge from the crossing over thus contain genetic material from both homologues: some of the original paternal material, and some of the maternal.

After crossing over, division continues much as in mitosis. In metaphase I the nuclear membrane breaks down and the bivalents – now mixed as a result of crossing over – line up along the centre of the spindle. Then comes anaphase I: the quartet of chromatids are pulled apart so as to form two sets of paired chromatids. Since each chromatid contains as much genetic material as a complete chromosome, the cell that results after anaphase I is still essentially diploid. It contains two complete sets of chromosomes – but the chromosomes just happen to be joined in pairs, at the centrosomes. Telophase then brings meiosis I to a close: the nuclear envelope starts to form again. Note in summary that the chromosomes do not reduce in number during meiosis I. The gametocytes are diploid when they enter meiosis I, and diploid when they emerge from it. The true task of meiosis I is to effect the genetic mixing.

Now follows a brief interphase. But this is not a true interphase. The chromosomes do not fully decondense. Neither – crucially – do they enter an S phase, and replicate. They begin to condense again even before they have fully decondensed. In short, while they are still diploid, the chromosomes embark on a second round of division – meiosis II. This, then, is when the reduction in chromosome number truly occurs. Meiosis I is about sex – the mixing of genetic material from the two homologous sets of chromosomes – and meiosis II is about reduction.

Prophase II is brief. The nuclear membrane disappears again; metaphase II begins; and now the chromatid pairs – still joined at the centromeres – are pulled apart in anaphase II. Once the chromatids are separated they can be called chromosomes. So after anaphase II we finally have four haploid sets of chromosomes. Telophase ensues, then cleavage of the cytoplasm, and we have not two but *four* daughter haploid cells.

Note that each gamete is genetically unique for two different reasons. First, when the doubled chromosomes divide at anaphase I, some maternally derived chromosomes and some paternally derived chromosomes travel to each pole. Thus, even without any crossing over, each gamete would be genetically quite distinct from its parent germ cell; and a human being, with twenty-three pairs of chromosomes to permutate, could in this way produce two to the twenty-third power (which is more than eight million) different haploid chromosome combinations. But secondly, since crossing over does occur as a matter of course, the number of different genetically unique combinations is effectively infinite. For this reason, then, every individual that is produced by sexual reproduction – by the fusion of two gametes – is also genetically unique. Cloning produces genetic duplicates and thus compromises this uniqueness. This, of course, is one reason that people object to it. If cloning were applied to human beings, we would object to it too.

The production both of sperm and of eggs involves meiosis; both go through the procedures outlined above. The gametes are, however, among the most specialised cells that animals produce, with the male and female versions designed for very different purposes and produced in very different ways.

The very different strategies of sperm and eggs

Thus the first task of gametes is first to convey genetic information into the next generation and the second but equally vital requirement is to provide each individual of the next generation with a food supply, until it can obtain nourishment for itself (which, in the case of mammals and some other animals including some sharks, may mean obtaining nourishment from the mother). However, gametes are also agents of sex – and sex implies the mixing of genetic information from different individuals. In order to meet this third requirement, then, gametes must be able to find and make contact with other gametes.

The simple solution would be for every individual in the population to produce the same kind of gametes – a simple, middle-of-the-road structure with a haploid set of genes, which contained a reasonable store of food, and was fairly mobile. In general, there would have to be some compromise between the need to carry a good food supply, to get the embryo off to a good start, and the need to be mobile, so as to be able to find another gamete. In fact many organisms, including some fungi, do all produce the same kind of gametes, and are said to be ‘isogamous’.

However, mathematical theory shows that isogamy is not the ideal course. If most of the gametes that the members of a population produce are middle-sized, compromise types, then natural selection will start to favour individuals who produce gametes that are slightly smaller than the norm, but are correspondingly more mobile and more numerous. These numerous and more mobile types will have a better chance of making contact with other gametes than the average type – although the resulting embryos will lose out somewhat because the smaller gamete will supply less nourishment. On the other hand, if some individuals in a population are producing smaller, more mobile gametes, then natural selection will favour other individuals who produce extra large gametes. These large gametes will inevitably be produced in smaller numbers and they will lack mobility. But they will carry extra supplies of food. Natural selection thus produces a runaway effect, with some individuals in a population tending to produce more and more gametes that are smaller and smaller and more mobile, while others tend to produce fewer and fewer and fewer gametes that lack mobility but are larger and larger. Hence sperm and eggs: sexual dimorphism at the level of the cell.

In practice, then, in animals and many other creatures as well, the sperm is as streamlined and cut down as it seems feasible to be. It has a head, which consists virtually entirely of a nucleus enclosed in membranes; a tail, which drives it along; a 'middle piece' between the head and tail, which contains a few mitochondria to power the tail; and an 'acrosome' at the front, a kind of nose-cone, which produces enzymes that enable it to penetrate the egg. The mature sperm contains almost no cytoplasm and supplies virtually zero nourishment to the embryo that it helps to form. It is a mobile packet of genes with a corrosive nose: a genetic missile.

Sperm are produced in the germinal layer of the testis from diploid spermatocytes; and each spermatocyte divides by meiosis to produce *four* haploid *spermatids*, which are little round cells that then mature into finished *spermatozoa* with heads and tails. The testis is laid down in the young fetus but the spermatocytes do not become functional, and start undergoing meiosis, until the young male reaches puberty; and then, at least in many species such as human beings, they go on turning out billions of spermatozoa every day virtually until death. In short, sperm production follows exactly the course you might anticipate: the gonad is established in the fetus, but does not become functional until sexual maturity; and each diploid spermatocyte gives rise, by meiosis, to four functional gametes.

But, although eggs are also produced by meiosis from diploid mother cells, their natural history is very different from that of the sperm and in several key respects is quite counterintuitive.

The peculiarities of eggs

As in males, the female gonads – the ovaries – with their germinal layers are laid down in the young fetus. The first surprise, however, is that they do *not* wait until the fetus is born, and becomes sexually mature, before they begin the process of egg production. The oocytes in the fetal ovary begin meiosis. They get as far as prophase I and then they 'arrest'. In fact in the species studied so far the first stage of meiosis seems to occur *only* in the young fetus. Once the fetal ovary has produced a good stock of oocytes in prophase I it never produces any more. This means that by the last third of its gestation a fetal mammal has *all* the eggs she will ever possess. Human baby girls are born with about three million eggs in their ovaries, and that is their lifetime's supply. The contrast with the male, whose testes do not even consider meiosis until the boy is mature and then function throughout life, is absolute.

The immature eggs remain in the ovary until the female is sexually mature and the eggs can be released from the ovary during 'ovulation'. The pattern of ovulation varies enormously from species to species but it is always 'cyclic'. That is, the female mammal ovulates (releases an egg) from the ovary, which passes into a passage called the '*fallopian tube*' and then down the egg duct or '*oviduct*'. If the female has successfully mated the egg confronts spermatozoa in the oviduct and is fertilised by one of them to become a zygote; in humans the egg remains receptive to sperm for about 20 hours after release. The zygote begins cell division as it continues to travel down the oviduct, cell division ensues, and finally the young embryo implants in the wall of the uterus. In humans and mice implantation occurs at the blastocyst stage; in sheep and cattle it is somewhat later. Hormones triggered by ovulation cause the wall of the womb (called the '*uterus*') to thicken, ready to receive the young embryo. If there is no pregnancy then the wall of the uterus thins down again, and the cycle must begin all over again. In women the whole cycle, from one ovulation to the next, takes around 28 days, in cows around 21 days, and in sheep around 16.

Women and cattle generally release only one egg per cycle and sheep commonly do so too, although many sheep (such as Finnish Landrace, which are highly prolific) may produce up to five oocytes at a time, and shepherds generally prefer twin lambs to singletons. Mice and pigs, of course, may produce a dozen or more eggs per ovulation. Women and many other animals produce an egg every month throughout the year but most species of mammal are seasonal breeders. Sheep, for example, undergo a few ovulatory cycles in the autumn and if they fail to become pregnant they simply stop cycling, and resume again in the following year. Since sheep pregnancy lasts five months an autumn mating should mean that the lambs are born in spring when the grass is growing: a summer of lush pasture to support lactation. Some domestic sheep, however, including the Dorset, come into season twice per year, in autumn and in spring. Sheep mate and conceive in late autumn and winter; which is why we do much of our serious work in winter (which can be mighty cold in Scotland) and why we have to plan our research season by season. In general, all female mammals are cyclic, but only some are seasonal.

The cycles in all mammals are controlled by roughly the same sets of hormones, operating in roughly the same ways, although the details do differ from species to species. But the seasonal ones have an extra tier of hormonal control: hormones from the brain, which typically respond to changing daylength. Outside the tropics, daylength rather than temperature is the best indicator of time of year. Thus Christmas Day in Britain may be balmy or it may be freezing, but it is guaranteed to be short.

Only a small proportion of the millions of eggs with which a female mammal is born mature and are released at ovulation. The rest just fade away as the years pass. Menopause ensues when the egg supply is running low, or indeed virtually exhausted; but only a few female mammals in the wild live to experience this. Elephants may live until they are 70 years or more although they rarely breed beyond their fifties. In general, though, a long period of post-ovulatory survival is a human luxury.

We take such cycles for granted; they are familiar enough, after all. But the details are strange. In the fetal ovary the eggs progress as far as prophase I – and then they arrest. They remain in prophase I until just before they are released from the ovary at ovulation or are simply resorbed (which is the fate of most oocytes). In fact, however, although the chromosomes are conventionally said to remain in prophase I they do decondense somewhat and can be seen through the electron microscope to take on the spread-out, 'lampbrush' structure that they have when their DNA is busy synthesising RNA. Thus the arrested oocyte chromosomes in prophase I look and behave like ordinary cells in the stage of interphase that is known as 'G2' (discussed in chapter 8).

Be that as it may, meiosis in the oocyte does not proceed beyond prophase I until the appropriate hormones have been released and chance decrees that one particular egg from among the many is to be selected for release. Then the favoured egg 'matures': that is, the oocyte proceeds from prophase I as far as metaphase II. In mouse eggs this maturation takes about 10 hours, in sheep and cow eggs around 24, and in pigs around 40. Thus the egg is in practice released in metaphase II (sometimes written as 'MeII' but generally more simply as 'MII') – when it is still diploid.

There are more peculiarities. As we have seen, meiosis in its theoretical, 'default' form should produce four haploid cells from one diploid mother cell. This indeed is what happens in the male: four perfectly functional spermatozoa are produced by every one dividing spermatocyte.

But the economics of egg production demand a modification.

The strategy of the egg is to provide the embryo with as much nourishment as possible. The cytoplasm of the oocyte is a cornucopia. Indeed the fully grown oocyte contains about 200 times more RNA than a typical body cell, about 60 times more protein, 1000 times as many ribosomes, and 100 times as many mitochondria in which to generate energy. It also builds up energy resources in the form of yolk, glycogen (a storage form of glucose) and lipids. To stock this larder it has sacrificed everything else; only a few eggs are produced (relative to the number of sperm) and mobility is abandoned. One of the evolutionary reasons why the oocyte remains in prophase I for so long is that in such a state it is tetraploid. Thus there are four sets of chromosomes per cell – four copies of every gene – and on the face of things this ought to enable it to produce more RNA and so create more proteins. Other 'helper' cells that surround the oocyte within the ovary, known generally as '*follicle cells*', also pile goodies into the oocyte cytoplasm.

The cytoplasm of the oocyte, in short, is a huge and precious investment. It would make no sense, having created such a storehouse, to break it into four during gametogenesis. It *is* necessary, however, to divide the original diploid genome into four. So egg production involves a neat compromise. Each of the two divisions of meiosis produces two daughter cells, each of which contains the same amount of nuclear material. Fair enough. But the division of cytoplasm is extremely unequal. Thus after meiosis I virtually all the cytoplasm finishes up in one daughter cell while the other daughter contains virtually no cytoplasm at all. The second daughter is just a little ball of surplus chromosomes, with a membrane, known as the 'polar body': a ping-pong ball against a football. The same thing happens in the second meiotic division: one huge daughter is produced weighed down with well-packed cytoplasm, plus another miserable polar body.

Thus, whereas spermatocytes produce four functional haploid spermatozoa, primary oocytes eventually produce one functional haploid egg plus two polar bodies. Of course, two polar bodies plus one egg makes only three daughters, and you would expect the primary oocyte to produce four daughters. But, in most species that have been looked at, the first polar body does not divide further. It simply sulks around and eventually gets lost. In mice, however, the first polar body may sometimes divide. In detail, species differ.

Finally, the mammalian egg at the time of ovulation is a more peculiar structure than most people – even most biologists – appreciate. It has a kind of a 'shell', known as the *zona pellucida*, which has various functions: exclusion of unsuitable sperm; mechanical protection and general maintenance of shape in the early embryo; and, in mammals, protection against a hostile immune response when the young embryo is travelling down the oviduct. The zona pellucida – often just called the 'zona' – is made from materials known as 'glycoproteins', which are compounded from carbohydrate and protein to produce a kind of tough plastic. The material in the zona is made partly by the oocyte itself, and in part is supplied by the supportive follicle cells that surround it in the ovary. There is a gap between the zona and the plasma membrane (the outer membrane) of the cell, known as the *vitelline space*.

The zona pellucida of a mammalian egg is not directly comparable with the shell of a bird's egg but in some ways the two are functionally similar. Thus early development in a mammal takes place within the zona. Indeed, when the mammalian embryo reaches the blastocyst stage it has to 'hatch' from the zona – 'hatch' is the technical term – before it can implant in the body wall of the uterus. Again, the idea that mammalian embryos must indeed hatch is alien even to most biologists. Elementary textbooks seem to contain no reference to it. Yet hatching is a key event in mammalian development. It is another nice reminder of the evolutionary past of mammals – for we are the descendants of reptiles, just as birds are.

So here we have a series of peculiarities. First, the egg remains in prophase I (or a form of it) from the time it is in the fetus to the time it is matured and released (assuming that it is released at all). Thus in long-lived creatures like women or elephants the developing egg will remain in prophase I for up to 50 years. Textbooks tend to present cell division as a brisk, no-nonsense affair, despatching its business in a few minutes. Well, mitosis in body cells can be rapid. But meiosis in eggs may take half a century to run its course. Secondly, despite what textbooks often imply and most people believe, the egg is released in a diploid state – as an MII oocyte – and is fertilised while still in that diploid state. As we will describe in later chapters, we have exploited this small detail to devise various contrasting methods for reconstructing embryos. Thirdly, the egg at the time of its release is surrounded by a shell, the zona pellucida, which the sperm has to penetrate; and the embryo develops within the zona and at some stage must hatch from it. People think of birds hatching, but not mammals.

If the female has mated at around the time of ovulation then her eggs should confront sperm in the oviduct. Then fertilisation can occur. Again, the details are pertinent to the technique of cloning by nuclear transfer.

Fertilisation and all that follows

Fertilisation has not one but *two* essential components – another very significant detail that is often overlooked. First, and most obviously, the entry of the sperm provides new genetic material – an entire *paternal genome* to complement the maternal genome that is already within the egg. But secondly, the sperm *activates* the oocyte. When the sperm first makes contact the oocyte is still diploid and indeed is still half-way through its second meiosis: it is in fact an MII oocyte. The oocyte has no nuclear membrane at this stage: the chromosomes are suspended within the cytoplasm, held in position by the spindle. The touch of the sperm on the oocyte's outer membrane stimulates the second meiosis to move to completion. The second polar body is then extruded, and becomes trapped in the vitelline space beneath the zona; and the remaining chromosomes acquire a new nuclear membrane and so for the first time form a haploid 'pronucleus'. Contact with the sperm also stimulates the cytoplasm to begin cell division – for, in the very young embryo, cell division is controlled by the cytoplasm.

The term 'activation' has two connotations. It implies the completion of meiosis II, and also the beginning of cleavage. The number of cell divisions in the early embryo that are controlled by the cytoplasm varies from species to species. In mammals, the cytoplasm controls only a few divisions, and then the genes of the new embryo take over. But in frogs the cytoplasm remains in charge until the embryo has around 4000 cells; and indeed frog egg cytoplasm, once activated, will happily cleave even after its nucleus is removed, to produce several thousand nucleus-less cells.

It is possible to prompt activation artificially: as we saw in chapter 4, Jacques Loeb achieved this in sea-urchin eggs simply by changing the chemical composition of the water. After such artificial activation the eggs of animals like sea-urchins may go on to develop into haploid embryos – that is, they undergo parthenogenesis. But, as we noted in chapter 3, parthenogenetic development in mammals cannot proceed far since mammals need both a maternal and a paternal genome. Yet the first part of activation – the completion of meiosis II – can be prompted artificially in mammals, for example by lowering the temperature in an oocyte that has been removed from its mother; and such activation, when accomplished inadvertently, very probably influenced and confused early attempts to clone animals by nuclear transfer. In the early 1990s we showed how the timing of oocyte activation affects the outcome of cloning by nuclear transfer, and in the experiments that led to Megan and Morag and beyond the scientists timed activation so as to maximise success. For now, though, just note in passing that activation – a process that non-specialists tend to forget altogether – seems to have a very significant influence on the success of nuclear transfer.

In practice, fertilisation requires tight coordination between the sperm and the oocyte. Lack of such coordination is one of several reasons why animals of different species may be unable to mate successfully, and indeed is one of many reasons for infertility even between males and females of the same species – including the human species. Thus the first contact between the head of the sperm and the zona induces the nose of the sperm to release its battery of enzymes which – aided by the thrust of the writhing tail – bores a track through the zona. Often, several sperm breach the zona; and then they lodge in the vitelline space, still swimming.

Except on rare and anomalous occasions, however, only one

of the sperm that has run the gauntlet of the zona sticks to the membrane of the oocyte itself. The outer region of the oocyte just beneath the plasma membrane is known as the *cortex*, and is studded with small structures known as *cortical granules*. As the sperm penetrates the zona the cortical granules release their contents, which in turn change the structure of the zona and prevent more sperm from entering. This release of cortical granules is part of the process of activation. The whole sperm enters the oocyte, tail and all; but soon the tail and the head membrane disappear, leaving only the small, highly concentrated nuclear material suspended within the oocyte cytoplasm. This nuclear material then expands to form a rounded paternal pronucleus, to complement the maternal pronucleus that should by now have formed.

So now at last we have two fully formed, haploid pronuclei, maternal and paternal, lying side by side in the cytoplasm. The whole resulting entity is a one-celled embryo, alias a zygote. What next?

Two genomes become one

Some books state that the two pronuclei in the zygote actually fuse, and this is what people tend to imagine if they think about such things at all. In fact, though, the two pronuclei do *not* fuse. Instead, when they get their breath back (metaphorically speaking) the chromosomes within each pronucleus enter an S phase, and divide. So now we have two *diploid* pronuclei sitting side by side within the zygote. At this point, the membranes around the pronuclei begin to break down, a spindle forms within the zygote cytoplasm, and the two sets of chromosomes from the two pronuclei line up compliantly along the axis of the spindle. In other words, mitosis begins. This is the first time that the maternal and paternal chromosomes come together. Mitosis then runs to completion – anaphase, then telophase, then cleavage – to produce a two-cell embryo. Each of the two cells now contains a diploid nucleus, containing both maternal and paternal chromosomes.

Note that the zygote – the one-celled nucleus – *at no stage* contains a single nucleus. First it has two haploid pronuclei, then two diploid pronuclei, and then it enters mitosis. There is no diploid nucleus, with a complete complement of chromosomes, until we reach the two-cell stage. This biological detail has all kinds of implications. For example, most people tend to assume that a new individual is 'conceived' when sperm and egg meet to create a zygote. But in the zygote the male and female genomes remain separate until the zygote itself divides. Do two divided individuals form an 'individual'? Then again, we cloned Dolly and the rest by transferring complete, diploid nuclei into enucleated oocytes. The result is a 'reconstructed embryo' with one cell. But this construct is not like the usual one-celled embryos that exist in nature; because one-celled embryos in nature do not contain single, diploid nuclei – unless they are formed as identical twins are: by splitting an embryo. This is a nicety, perhaps with no practical consequences. But you never know when details of this kind may turn out to be important, as the history of cloning shows.

One last technical point of serendipity. The first polar body (produced at meiosis I) inevitably remains trapped in the vitelline space, and sits just next to the nucleus of the egg itself: the egg nucleus just beneath the egg cell membrane, and the polar body just outside. The cytoplasm of an oocyte is opaque (at least in hoofed animals like sheep) and for us (generally Keith or Bill Ritchie) whose task it is to remove the genetic material from the egg in the course of cloning, the first polar body provides an indispensable marker. Enucleation is in practice achieved by locating the polar body (it is clear to see as a dark blob in the area beneath the zona) and then sucking out the cytoplasm that lies immediately beneath it. Usually this sucked-out cytoplasm will contain the egg nucleus. Now, new ways of locating the underlying chromosomes are being developed based on location of the spindles. But, before these new techniques existed, it is hard to see how we could have removed the oocyte chromosomes at all if we hadn't had the polar body to act as guide. Note in passing, too, that the term 'enucleation' is not strictly appropriate when applied to the MII oocyte, since MII oocytes do not contain properly formed nuclei. The nuclear membrane is absent in an MII oocyte, and the chromosomes lie within the cytoplasm, held in place by the spindle. But 'enucleation' is a convenient term.

But that is enough background. Now we can begin Part III of this book and look sensibly at the cloning of mammals – the work that led from Megan and Morag through Taffy and Tweed and the famous Dolly to the *pièce de résistance*: Polly.

PART III

The Path to Dolly

CHAPTER 6

Mammals Cloned

Ian Wilmut:

In the last three chapters we have outlined the essential biological ideas that lie behind our kind of cloning. Most of those ideas were developed by other scientists, many of whom lived in a quite different age, with a quite different idea of how life works, but they nevertheless founded the science on which we have built. So now we can resume our own story, and describe what we, Keith and I, have done ourselves.

I suppose I should trace my own involvement back to the early 1970s, when I worked at Chris Polge's lab at Cambridge and produced Frostie, the first calf in the world ever to be born successfully from a deep-frozen embryo. But, as described in chapter 2, the sequence of research that eventually produced Megan and Morag, Dolly, and Polly, really began back in the early 1980s when – somewhat unhappily – I found myself recovering sheep embryos from donor ewes, then adding DNA, by what was then (and indeed in most contexts still is) the standard method of genetic engineering. This work was very fruitful – Tracy came out of it – but I knew from the outset that there must be much more efficient ways to effect genetic transformations.

I would like to say that I knew from the beginning what the route would be. I would like to say that I perceived, in the first year if not the first week, that we should begin by culturing sheep cells, then transform them genetically while they were in culture, and then clone entire new animals from the cells that had been transformed most satisfactorily. It would be good to claim that I foresaw from the beginning the conceptual path that was to lead through Megan and Morag to Dolly and then to Polly – but that would be a complete fantasy. The fact is that in the early 1980s no one knew for certain that mammals could be cloned at all by nuclear transfer. John Gurdon had cloned frogs by this means, but mammals are not frogs. Certainly, no one had thought of cloning whole new animals from cultured mammalian cells. Certainly not from cultured *adult* cells. All these ideas emerged during the 1980s and early 90s, and they came as a series of revelations.

In fact the rate of development in mammalian cloning – from not knowing whether it was possible at all in the early 1980s, up to Megan and Morag, Dolly, and Polly in the mid 1990s – is astonishing. There have been comparable bursts of speed in the history of science and technologies: penicillin was taken from a laboratory curiosity to an invaluable drug within a couple of decades, and NASA put a man on the Moon within 15 years or so of their first, modest, orbiting Earth satellite. In those cases, though, the scientists and technologists involved knew where they were going from the start and had a good idea of what they needed to achieve along the way. We who have followed the path of cloning – not just at Roslin, but also at other labs in Britain, Europe, and the US – had very little idea from year to year what could actually be done. We have followed our noses. Along the way, too, there have been diversions and noises off: happenings of the kind that are rare in science, and have become *causes célèbres*. I hope that the account that follows will seem straightforward – but it is written with the twenty–twenty vision of hindsight. Much of the time in the brief history of modern cloning – from the 1970s to the 1990s – it was impossible for anyone, even for those in the field, to know what was actually true, or what to make of all the conflicting information. Megan and Morag and Dolly would be extraordinary even if we had known from the outset what we were trying to achieve, and what was actually possible. The way things actually happened, with all the side-tracks and misconceptions, makes them doubly remarkable. But then, perhaps their story is typical – for although science progresses it rarely proceeds in straight lines. Science is, after all, an attempt to explore the unknown. And if you venture into unknown territory, how can you avoid getting lost?

Ian Wilmut and Keith Campbell:

We can usefully begin the modern story of cloning in the 1970s – a productive decade, but also highly confusing. Many of the necessary techniques and ideas were developed or at least initiated then. Young mouse embryos were developed *in vitro*, and embryo transfers were carried out in mice and cattle in various labs – including Frostie, at Cambridge.

Nuclear transfer was performed, too, in mice and other species including – in Derek Bromhall's excellent studies at Oxford – rabbits. Some of these transfers required enucleation of the recipient cells, and some did not; in the latter, nuclei from other cells were simply introduced into oocytes or zygotes without first removing their nuclei. Such studies were in general intended to investigate the dialogue between nucleus and cytoplasm; the switching on of the genes in the early embryo (of which more later) and during differentiation in general; and the phenomenon of totipotency, and its loss during differentiation. Steen Willadsen at the ARC's Institute of Animal Physiology at Cambridge announced the first clones of sheep in 1979. He did not do this by nuclear transfer, however, but simply by splitting a two-cell embryo – but along the way he developed methods for protecting young sheep embryos which we have incorporated into our routine at Roslin. More of this later, too.

But, despite these advances, the future remained unclear. Indeed through the 1970s and well into the 80s there were reasons to doubt whether mammals could be cloned by nuclear transfer at all. There were also diversions, including works of fiction. Most of these, like Ira Levin's *Boys from Brazil*, which describes the cloning of Nazis deep in the jungle, were presented as conventional novels and led nobody astray. But David Rorvik wrote a work of fiction about the cloning of an eccentric millionaire which he claimed was *fact*, and it was published as such in 1978 by J. B. Lippincott of New York with the title *In His Image: The Cloning of a Man*. For a time, some people – including some biologists – believed what Rorvik said. After all, he gave his work a veneer of authenticity by citing Derek Bromhall's work (although Bromhall later sued the publishers for dragging his name in).

Overall, then, developmental biology had a great deal more expertise and insight at the end of the 1970s than at the beginning, but the particular issue of mammalian cloning by nuclear transfer was still up in the air. Had the necessary techniques already been developed, as Rorvik suggested? Few biologists really believed that – yet it was almost plausible. On the other hand – was mammalian cloning by nuclear transfer possible at all? Was there something peculiar about mammals, and in particular about differentiation in mammals, that would prevent this? Or did the truth lie somewhere in between – that cloning by nuclear transfer was possible, but only up to a point, and only when a great deal more groundwork had been done? In 1980, no one could answer these questions. In the 1700s, when science itself was new, most subjects were in this precarious state. In the modern age, such deep uncertainty was rather strange.

But as the 1980s began it seemed as if all the uncertainty had been resolved at a stroke. Or at least: it seemed that way for a short time. For mammalian cloning seemed to get away to a flying start in the early 1980s at the hands of Karl Illmensee (yet another German embryologist), and Peter Hoppe, an American. Illmensee was widely credited with genius. He had been impressing people since the 1960s when he produced an excellent PhD thesis on the genetics of fruit flies; and the very accomplished Hoppe was the perfect foil. Together they amazed everyone in the late 1970s by announcing that they had produced baby mice with all-male or all-female parents: that is, by adding a female pronucleus to an unfertilised egg, or removing the female pronucleus and adding two male pronuclei. As we will see, biologists later began to perceive that mammals will not develop successfully unless they inherit genomes from both sexes; but in the 70s Illmensee and Hoppe's research was taken at face value and, very reasonably, was thought to be both salutary and brilliant. Illmensee became a professor at the University of Geneva and Hoppe went with him. They were an outstanding team.

Their paper on the cloning of mice in *Cell* in January 1981 seemed to solve most of the problems in one stroke. They announced the birth and subsequent good health of three cloned mice, created by nuclear transfer. Furthermore, they had taken the donor nuclei – '*karyoplasts*' – from relatively advanced embryos. Thus they also provided the means for doing this: microsurgery, by pipette. For a time it seemed that mammalian cloning was all over bar the shouting. All that was needed, apparently, was good technique.

The *Cell* paper is a model of clarity. Mice, said Illmensee and Hoppe, are the ideal creatures in which to study cloning. For cloning is a problem both in development – how to manipulate cells without killing them; and in genetics – for development involves the differentiation of cells which (as was clear by the 1980s) requires the genes of the genetic programme to come on line sequentially. Amphibians are excellent for developmental studies – all those big, robust eggs – but their genetics is only poorly understood, and since they have long life cycles (they do not breed until they are two or three years old) it would take an enormous amount of time to study their patterns of inheritance and hence to gain insight into their genes. Fruit flies have much shorter generation times and a simple genome and have been geneticists' species of choice since the pioneering studies of the great Thomas Morgan in the 1920s. But fruit flies belong to the group called 'protostomes', which develop very differently from vertebrates (although studies in the 1990s have shown more similarities than expected!). For studies that involve development *and* genetics, and throw light on vertebrates in general and mammals in particular, mice seem ideal. They have short generation intervals – since they are fertile at ten weeks they are nearer to fruit flies than to frogs – and their genetics is well understood, while they have long been accepted as the natural 'models' by which to study mammalian development and physiology.

The experiments as reported in *Cell* were extraordinarily bold.

Hoppe and Illmensee took their karyoplasts from blastocysts that had already differentiated into the two main groups of cells: the inner cell mass (ICM), which develops into the embryo itself, and the surrounding trophectoderm (TE), which at the start of pregnancy makes the initial contact with the wall of the uterus, and so provides the tissues that precede the placenta. They thought that if the karyoplasts contained too much cytoplasm then this would compromise the reconstructed embryo. So they sucked nuclei both from ICM and from TE cells with a pipette that had a tip that was smaller than the cells – around 10 microns – which extracted the nuclei together with just some of the cytoplasm, while stripping off the outer cell membrane.

Their method for enucleating the recipient cells (the enucleated cells are termed '*cytoplasts*'), and introducing the karyoplasts, was similarly neat and ingenious. A method for fusing cells had been developed in the 1960s, using Sendai virus. The virus breaks down the outer membranes of the cells – but only temporarily; and as the membranes re-form so they fuse and thus enclose the two cells together. In fact, many scientists continued to use Sendai virus for cell fusion at least until the mid 1980s but Hoppe and Illmensee rejected this method – for, they said, attempts in the 1970s to reconstruct embryos by fusing with the virus 'resulted at best in a few cleavage divisions'. Perhaps, they suggested, the virus itself has untoward if unknown side-effects.

Instead of fusing by virus, they again opted for microsurgery – effectively, as they acknowledge, adapting the technique described by Derek Bromhall in rabbits in the 1970s. First they injected the karyoplast directly into the cytoplast, using the same pipette that they had employed to extract the karyoplast in the first place. Then they sucked the cytoplast nucleus into the pipette, and withdrew the pipette with the cytoplast nucleus inside. Thus they needed to stab the pipette into the cytoplast only once, and so made only one wound. If they had removed the cytoplast nucleus first, which seems the more obvious thing to do, they would have needed two stabs. Secondly, they made use of yet another of those serendipitous agents that make experimental biology possible: cytochalasin B, produced by a mould. Cytochalasin B inhibits the replication of the protein molecules that form the cytoskeleton: the tough, fibrous network that supports the cytoplasm. In other words, it softens the cytoplasm. It is now standard to treat the cytoplast with cytochalasin B before fusion – this has become routine with us at Roslin – but in 1981 the technique was still novel. With cytochalasin B, said Illmensee and Hoppe, 'the survival rate of injected eggs improved significantly'.

As nuclear donors, Illmensee and Hoppe used two different strains of mice: a grey one (strain LT), and one with a brown speckly coat of the kind known as 'agouti' (strain CBA). The recipient cytoplasts were enucleated zygotes (note zygotes – one-celled embryos, not oocytes – diploid eggs) taken from black mice (of strain C57B1). It is important to use donors and recipients of different strains; then you can see whether any offspring that result truly inherited their genes from the donor nucleus, or from the recipient's DNA, inadvertently left behind at enucleation. Partly for this reason, when we made Dolly, we transplanted cells from a Finn-Dorset ewe into a Scottish Blackface oocyte. The only difference is that breeds of sheep have lovely, folksy names that have grown out of farming, while strains of laboratory mice are known by ugly codes. Nowadays it is possible (and necessary) to check the genetic provenance of all cloned creatures by 'genetic fingerprinting', of which more later; but that technique was not available in the early 1980s.

To use both ICM and TE cells as karyoplasts was a touch of class. It was already suspected that these two layers were differentiated to different degrees. Thus, embryologists had already shown that ICM cells happily integrated themselves when transferred into another embryo, and developed into appropriate tissues. As Illmensee and Hoppe put it, ICM cells retain 'remarkable developmental plasticity'. But TE cells, when transferred from embryo to embryo, are much less amenable to change.

Illmensee and Hoppe cultured their reconstructed embryos for about four days – keeping them alive and allowing them to develop *in vitro*: without the comfort of an oviduct, where they would normally be residing at this stage. Already it was clear, however, that the embryos that began with ICM nuclei and those with TE nuclei were very different. Out of 179 embryos reconstructed with TE cells, sixty-eight survived the microsurgery but of those only seven became morulae. Most of those morulae had abnormal cells – of the wrong size or density – and only one of them became a blastocyst. Illmensee and Hoppe concluded that nuclei derived from TE cells will not generally produce viable embryos.

But the 363 embryos constructed from ICM nuclei fared very differently. True, only 142 survived the initial microsurgery – less than forty per cent – but forty-eight of those 142 developed through the four days of culture to become normal morulae or blastocysts. For this kind of experiment, particularly in its pioneer phase, this is a commendable haul.

Illmensee and Hoppe then selected just sixteen of the ICM-derived morulae and blastocysts and transferred them into the uteri of five of the white Swiss mice. Whatever the circumstances, it is standard practice in embryo transfer to introduce several embryos at a time. Doctors do this in human IVF (*in vitro* fertilisation) clinics. In our experiments with Megan and Morag and Dolly, we placed two young embryos into each receiving ewe. It is as if the surrogate mother needs to be reminded firmly that she has embryos on board, seeking refuge in the uterine wall. Indeed, at the same time as Illmensee and Hoppe introduced the LT and CBA mice, they also included more embryos from yet more strains (BALB/c and Swiss). These extra embryos served two functions: first they helped the pregnancy along; and secondly they were just ordinary embryos – *not* reconstructed – and so they acted as controls with which to compare the reconstructed embryos.

Anyway, of the sixteen reconstructed embryos that were transferred into surrogate mothers, no fewer than three went to term and were born alive: two females and one male. Mice are born naked but two of them – a male and a female – went on to acquire the grey coats typical of the LT strain, while the remaining female developed the typical agouti coat of a CBA. Three live births out of sixteen transfers is 18.8 per cent. The unreconstructed embryos who were transferred alongside them and became their littermates fared much better: more than seventy per cent of them went to term. But the difference hardly matters.

That *any* of the reconstructed embryos came through at all was wonderful.

When two of the reconstructed LT mice were ten weeks old Illmensee and Hoppe mated them with normal animals of their own breed. By the time they submitted their paper to *Cell*, the male nuclear transplant mouse had sired sixty-two offspring, and the female had given birth to twenty-seven: all with the typical grey coats of the LT strain. The two LT transplant mice were also mated with each other, and together produced twenty-four normal offspring. Truly the ICM nuclei that gave rise to them must have been totipotent.

Illmensee and Hoppe concluded their astonishing report in *Cell* with recommendations for further research. Thus, they said, 'In future experiments it will be important to determine in more detail to what extent the transplanted nucleus becomes regulated by the egg cytoplasm,' and, 'It will be important to reveal the nuclear potential of various cell types at more advanced embryonic and fetal stages ... to gain more insight into the biological consequences of genomic modifications during mammalian cell differentiation.' Yes indeed. Illmensee and Hoppe were outlining, in the early 1980s, the precise research programme that others have followed since, including us, albeit in sheep rather than in mice.

So what more was there to say? Mammals had been cloned by nuclear transfer, just as John Gurdon had cloned frogs. Furthermore, they had been cloned from ICM cells, which, though not apparently very differentiated, are a lot further advanced than the blastomeres of two- or eight-cell embryos. For good measure, Illmensee and Hoppe had shown a clear difference in potential between ICM cells and TE cells. They had also developed neat and clean techniques for enucleation and for cell fusion: microsurgery by pipette, aided by cytochalasin B to soften the cytoskeleton. Brilliant. Dazzling, even. A pre-emptive stroke.

But all was not so simple. The essence of science is repeatability. Trivial experiments in science that demonstrate nothing very much are not necessarily repeated – life is too short – but anything as momentous as this has to be done again. But other scientists could not reproduce Illmensee and Hoppe's results. Although Illmensee insisted that it was just a matter of technique he seemed reluctant, in practice, to demonstrate his methods to colleagues (although Hoppe had no such reservations). Eventually, yet another embryologist in the great German tradition, Davor Solter, together with his American colleague James McGrath, set out to repeat Illmensee and Hoppe's 1981 experiment more or less precisely. Their results were negative: so much so they effectively ended Illmensee's research career and severely dented the confidence of many other biologists who were seeking to clone mammals.

By the early 1980s Davor Solter and James McGrath, at the Wistar Institute of Anatomy and Biology in Philadelphia, had also established fine reputations in developmental biology and introduced significant innovations in nuclear transfer. For example in 1983 in *Science* they described a method for enucleating a mouse zygote with a pipette *without* penetrating the outer membrane from the outside at all. They simply pushed the tip of the pipette through the zona pellucida and placed it against the cell membrane adjacent to the pronuclei which lay just beneath, and sucked them out one by one. This is yet another technique that we have borrowed for our own work at Roslin. When it came to fusing karyoplast with cytoplast, Solter and McGrath generally preferred to employ Sendai virus, rather than direct injection. Thus they achieved nuclear transfer without ever penetrating karyoplast or cytoplast at all.

With such small modifications of technique, McGrath and Solter set out in essence to repeat the experiments that Illmensee and Hoppe reported in 1981; reconstructing embryos of mice, and then culturing them to the blastocyst stage. But where Illmensee and Hoppe described one bold foray with ICM cells, McGrath and Solter undertook a step-by-step series of experiments, transferring karyoplasts of more and more advanced stages into cytoplasts that were also derived from embryos of various stages. This is the kind of methodical approach that I (Ian) tend to favour: I carried out comparable studies in sheep in the late 1980s. In general, McGrath and Solter's transfers were successful when the karyoplasts came from very young embryos, and/or when the karyoplast and cytoplast came from embryos of a similar age: but the transfers failed when the karyoplasts were older, and when karyoplast and cytoplast were at different stages. McGrath and Solter could not do what Illmensee and Hoppe claimed to have done. Embryos constructed from ICM karyoplasts simply perished. They conveyed their negative result in the less-than-snappy title of their paper in *Science* on 14 December 1984: 'Instability of mouse blastomere nuclei transferred to enucleated zygotes to support development *in vitro*.'

McGrath and Solter's methodical approach – transferring karyoplasts from embryos of steadily increasing age – is vital, for two reasons. First, if nuclei from later-stage cells did indeed lose their potency, then they wanted to see when exactly this happened. But secondly, and more broadly, it is essential to establish whether the experimental failures result from bad technique, or because the system under study is innately incapable of doing what is asked of it. If McGrath and Solter had simply transferred ICM nuclei into enucleated zygotes, and these had failed, they would not know whether this was because ICM nuclei really could not support development or because their technique was wanting. But if they first showed that their technique succeeded when the nuclei came from very young embryos, then they would know that any failure with ICM karyoplasts resulted from the nature of the cells, and not from the method of transfer.

So McGrath and Solter began at the beginning and first transferred nuclei from one zygote to another: enucleating the recipient by their sucking method, and fusing karyoplast and cytoplast by Sendai virus. This worked perfectly. When the resulting, reconstructed zygotes were cultured, they said, 'Nearly all developed to the blastocyst stage.' Clearly, there was nothing very wrong with their technique.

So now they fused karyoplasts taken from two-cell embryos with enucleated zygotes. Now only nineteen per cent of the reconstructed embryos developed to the morula or blastocyst stage. Four out of five perished. That a minority did succeed, however, again seemed to vindicate their technique.

Then they moved up to four-cell embryos – and now the results were dismal. Only four out of eighty-one embryos that they constructed using karyoplasts from four-cell embryos developed beyond the two-cell stage – and as these four developed into morulae, they all showed abnormalities. Then they took karyoplasts from eight-cell embryos – and the results were even worse. Only one out of 111 of these reconstructed embryos developed past the two-cell stage – but then it petered out as the cells tried to divide again.

Finally, McGrath and Solter did as Illmensee and Hoppe had described, and transferred nuclei from ICM cells into zygotes. Only one out of eighty-four of the embryos that they constructed with ICM cells got past the two-cell stage, and that perished at the four-cell stage. Yet these embryos were precisely the kind that Illmensee and Hoppe said had sometimes survived to become adult mice.

McGrath and Solter then analysed their own technique, step by step, to see if it did have shortcomings after all. Perhaps they were treating the embryos too roughly. Perhaps the medium was at fault. Perhaps the Sendai virus used for cell fusion was damaging. Perhaps they should not expose the embryos to room temperatures. So they subjected a series of embryos to each of the possible insults one by one but *without* nuclear transfer – surgery, medium, virus, cooling – and ninety per cent of the embryos that were thus treated went on to develop into morulae or blastocysts. The techniques taken alone, then, seemed benign enough.

It seemed difficult to escape the conclusion, then, that the failures should be laid at the door of the transferred karyoplasts. As they got older, they indeed lost their potency: their power to support development to the blastocyst stage and perhaps beyond. But then, a karyoplast is not just a nucleus. It is nucleus plus cytoplasm. Should the shortcomings of the older karyoplasts be ascribed to their nuclei or to their cytoplasm?

So now came another series of experiments. McGrath and Solter fused mouse zygotes that had *not* been enucleated with cytoplasm taken from cells of different ages – a straight reversal of normal procedure, where the zygote is enucleated and the donor cell retains its nucleus. Embryos constructed in such ways nearly always developed happily into blastocysts. Clearly, the cytoplasm of transferred karyoplasts is not to blame for any failures of development.

But then they fused entire karyoplasts (still with their nuclei) with entire zygotes (zygotes that again had not been enucleated) and so produced tetraploid embryos. This time the results were very different. When the karyoplasts came from two-cell embryos, seventy-five per cent of the reconstructed embryos developed to the blastocyst or morula stage. But when the karyoplasts came from four-cell embryos, only fifty per cent of the reconstructed embryos reached morula or blastocyst. When the karyoplasts came from embryos that were eight-celled or above, less than one per cent of the reconstructed, tetraploid embryos went on to become morulae or blastocysts. In short, nuclei from older embryos not only failed to support development when placed in enucleated cytoplasts; they also upset the development of zygotes that retained their own nuclei.

Still McGrath and Solter persisted. They really are the most tenacious investigators. Is it possible, they asked, that nuclei from two-cell embryo are simply less robust than nuclei from zygotes? Perhaps they don't really lose potency at this age. Perhaps they simply become less resistant to transfer. What would happen if they transferred nuclei from two-cell embryos into enucleated cytoplasts taken from two-cell embryos? In such cases, after all, the nuclei from two-cell embryos would not be expected to re-enact the role of a zygote nucleus. They could simply carry on as normal, in the environment of a two-cell embryo. If they failed to do so, it would show that they had simply lost resilience. Twenty-two times, then, McGrath and Solter constructed embryos by fusing karyoplasts from two-cell embryos with enucleated cytoplasts from two-cell embryos. Eighteen of the twenty-two went on to become blastocysts. In short: older nuclei do not object to transfer *per se*. They merely object to being put into zygote cytoplasm. They are not enfeebled. But they are not totipotent any more.

Finally, McGrath and Solter asked again whether their method of fusion – using Sendai virus – was at fault after all. So they set out to repeat Illmensee and Hoppe's procedure to the letter. They injected ICM cell nuclei into zygotes, and then removed the pronuclei of those zygotes. About a third of the embryos survived the microsurgery and about forty per cent of those survivors divided once, while a few divided twice. But, they say, 'No embryo developed beyond the four-cell stage.' Still, they asked, was the procedure too rough? So they injected culture medium into control zygotes instead of ICM karyoplasts to simulate the injection; and they removed cytoplasm from other controls to simulate the rigours of enucleation. These control zygotes survived the abuse and developed perfectly well. So at last McGrath and Solter concluded: 'ICM cell nuclei are unable to support development with enucleated zygotes.' These conclusions, they added, 'contrast with the results of Illmensee and Hoppe'.

Then McGrath and Solter went on to make a statement that resounded through the world of developmental biology – although it has somewhat haunted them since: 'the cloning of mammals, by simple nuclear transfer, is biologically impossible.' Authority matters in science, although in principle it should not: and McGrath and Solter were, and are, authorities, and their words had impact. Yet they were clearly wrong – or at least, it depends what you mean by 'simple' nuclear transfer. Within a few years of their all-too-gloomy asseveration other mammalian species *had* been cloned: sheep, cattle, goats. As Keith says, 'Some things really are impossible, but others are merely difficult.'

But how had Illmensee and Hoppe apparently succeeded while McGrath and Solter failed with exactly the same procedure? The answer is not clear. Illmensee certainly was and is a brilliant experimenter; but then, so are McGrath and Solter. Keith says: 'I believe that Illmensee and Hoppe might well have managed to produce live mice from ICM karyoplasts (and, as we will see, in 1998 Japanese biologists produced adult mice from *adult* cells – the murine counterpart of Dolly). They could not have known exactly how they succeeded – the necessary theory and techniques were not developed in the early 1980s – but they could have got lucky. These things happen. I am also grateful to Illmensee. I attended a lecture of Illmensee's in 1984 at Sussex, when I was just beginning my DPhil, and he was certainly inspiring. Who knows, without him, I might not have got involved in this field at all!'

Many, though, felt that Illmensee and Hoppe had claimed more than they could justify. The University of Geneva appointed an international commission to investigate their work, which met in August 1983. The committee did not conclude that Illmensee and Hoppe had fabricated results but they did criticise the way they had written them up and declared that their work should be repeated 'with full scientific rigour' – which indeed, is what McGrath and Solter did. Illmensee resigned from the university in July 1985 when his contract ran out (Swiss professorships run only for limited terms) and now works in Salzburg on human IVF. Peter Hoppe was never directly criticised but retired nevertheless in 1985. Modern papers on cloning sometimes cite Illmensee's work, and sometimes do not. Many feel in their bones that his work is important, whatever the details; and so it surely is. The whole episode is a terrible pity in many different ways; and among them we may simply say, 'What a waste of talent!'

Thus the 1970s produced some excellent advances but also noises off – including works of fiction – that muddied the waters; and then at the start of the 80s the waters were muddied again – not by noises off, but by events within science itself. One minute a scientist with an enviable reputation claimed virtually to have solved the problem of mammalian cloning and the next, scientists of equal stature stated that such cloning is 'impossible'. Hindsight shows, too, that for such experiments mice are rather a bad choice. To be sure they have often proved to be excellent laboratory 'models' in which to study the ways of mammals in general but as things have turned out they may be *more* difficult to clone than sheep and cattle (although, as we will see in chapter 9, the techniques developed for Megan and Morag at Roslin do seem to overcome the problems with mice as well).

With hindsight, too, we can immediately identify one obvious reason why both Illmensee and Hoppe and McGrath and Solter were liable to fail, especially when they used ICM cells as karyoplasts. Both transferred nuclei into zygotes – whereas, as will become apparent from chapter 8 onwards, if you really want to succeed you need to use MII oocytes as cytoplasts. At least, this is true if the genomes of the karyoplasts have lost some totipotency, and need to be reprogrammed, because MII oocytes (if taken at the right time) contain the factors needed for reprogramming, and zygotes do not. By the zygote stage, apparently, those factors have already been used up. But again we are running ahead of ourselves. We will come to all this.

But in 1984 it was not evident to McGrath and Solter that oocytes are preferable to zygotes and in their 1984 paper they suggested three other reasons why their attempts at cloning had failed. Their first concerns what has been called 'the maternal to zygotic transition', or, more succinctly, 'genome activation', which clearly does matter in this context. Their second is the phenomenon of genomic imprinting, which is extremely important in general, but may not be so relevant in this context (although the jury is still out on this). Their third suggestion is that the cell cycle is important and may need to be controlled. The cell cycle is so significant it deserves a chapter to itself (chapter 8); indeed it is the key to the whole thing. But first we should look at the other two phenomena that McGrath and Solter identified, one of which certainly matters and the other may well turn out to be important in cloning, even though its relevance if any is not yet obvious.

Genome activation

The DNA in the nucleus normally does two things during the life of a cell. It replicates itself, making two DNA molecules where before there was one; and it makes protein. In order to make protein, DNA first has to make RNA; this RNA creation uses DNA as a template (transcription). All this was outlined in chapter 2.

The DNA in the cells of the early embryo replicates like mad as the zygote becomes a two-cell embryo and the two cells become four cells, and so on and so on. But, in most animals, the DNA in the very early embryo is *not* transcribed. It therefore produces no protein. All the enzymes needed to assist DNA replication, all the extra proteins needed to create more cell membrane as the cytoplasm divides, all the metabolic cycles needed to provide the energy to drive the whole process, and most or all of the instructions that direct the whole process of cell division, are provided by the cytoplasm of the zygote and then by cells of the young dividing embryo. The cytoplasm of the zygote and its daughter cells originally derived from the oocyte that was initially fertilised; and this oocyte cytoplasm was originally built within the ovary, with help from various other cells. In short, the DNA in the early embryo effectively does what it is told. As we have seen, there is no growth in the early embryo – none until we get to the blastocyst and beyond. DNA replication takes place without cell growth. The successive generations of cells just become smaller and smaller, until they are down to the size that is typical of the species in question.

Sooner or later, however, the DNA of the young embryo does become active. Transcription begins. The genes are 'switched on'. Then they do assume the driving seat. This crucial phase in the life of the early embryo has been called 'the maternal to zygote transition'; but since it does not generally take place in the zygote (the one-cell embryo) it is perhaps better known simply as *genomic activation*.

Developmental biologists can observe genomic activation taking place, more or less directly. For example, RNA contains the base known as uracil, whereas DNA contains thymine instead. If you add radioactive-labelled uracil to young embryos in culture you can see at what stage they start to incorporate it. They incorporate the uracil only when they start to make RNA – which, of course, signals the start of transcription. There are secondary signs, too, of genomic activation. Notably, the interval between cell divisions (i.e. the length of the cell cycle) tends to increase as the genes come on line – for the simple reason that transcription, and protein production, are time consuming.

By this and other means during the 1970s and 80s biologists were able to identify virtually the precise point at which the genes in the young embryo are switched on. Of the creatures studied, frogs have the laziest genes. They do not become active until the embryo contains 3000 to 4000 cells. Indeed you can remove the nucleus from a frog zygote and, with suitable stimulation, its cytoplasm will still divide to form an embryo of several thousand cells – though all of them lack a nucleus. Here is demonstration indeed that at this stage the cytoplasm is in charge. The nuclei seem virtual passengers.

In mammals, however, genomic activation occurs sooner. In pigs and rats the DNA begins to be transcribed at the four- to eight-cell stage. In cattle and human beings the DNA comes on line at the eight-cell stage. In sheep there are signs of transcription at the eight-cell stage although the genome is not properly active until the sixteen-cell stage.

But in mice there are signs of genomic activity while the embryo is still a zygote: even before the first cell division has taken place. At the two-cell stage the mouse genome is clearly active.

Why does genomic activation affect the possibility of cloning? Well, in fine detail, the answer to this is still not known (and with modern techniques, of the kind that produced Megan and Morag, it apparently ceases to be absolutely critical). But it is easy to envisage in principle why genomic activation matters. To replicate animals by cloning, a nucleus from an embryo cell – or indeed from older tissues – is placed in the cytoplasm of a zygote or oocyte, and is then expected to manifest all the totipotency of zygote pronuclei, able to give rise to successive generations of cells which between them will express all the genes appropriate to all the tissues of the species. If the genome in the transferred nucleus is from some specialist cell and if totipotency is to be restored, then the genome must be 'reprogrammed'. Intuitively we might guess that the less differentiated the transferred nucleus is, the easier the reprogramming would be; this after all is what John Gurdon showed in frogs in the 1960s. A genome that has not even been switched on, and has not even begun to transcribe, surely would be easiest of all to reprogramme. Thus we might envisage that increasing differentiation would present a steadily rising slope of difficulty, while the particular event of genomic activation is a specific barrier that has to be overcome (and may make itself felt even before any differentiation is discernible). As we will see, the notion that totipotency is indeed reduced as differentiation progresses inspired much of Ian's work on cloning through the late 1980s and into the 90s.

McGrath and Solter point out in their *Science* paper of 1984 that cloning of frogs is probably made easier because genomic activation is delayed for a very long time while 'the embryonic genome in the mouse becomes active in the two-cell stage, or even in the zygote'. Perhaps, though, having perceived that there are such differences, they should not have been so ready to extrapolate from mice to other species and to suggest that 'the cloning of *mammals*, by simple nuclear transfer, is biologically impossible'. They might have considered that cattle and sheep would be easier since genomic activation is held up at least for a few generations of cells. So it has proved.

In their 1984 paper McGrath and Solter also drew attention to genomic imprinting. So far, this phenomenon does not seem to raise particular difficulties in cloning. But that may be an illusion; as Ian points out, 'The success rate in cloning is still extremely low – and how do you know genomic imprinting is not contributing to the failures?' It might also cause problems later in life – which cannot be known as yet as Dolly is still young. We should look also at genomic imprinting.

Genomic imprinting

In the 1860s Gregor Mendel did, alone and if anything hindered by his few *confidants*, establish the main principles of genetics with his experiments on peas and other homely plants in his monastery garden at Brno in what is now the Czech Republic. He was one of the great geniuses of all science.

But he did get one thing ever so slightly wrong. In his seminal paper of 1866 he stresses that it makes no difference at all whether an organism (a garden pea, in his experiments) inherits a particular character from its mother or from its father. The 'hereditary factor' (his name for genes) for yellow flowers dominates the factor for 'white' flowers, so when yellow-flowered peas are crossed with white-flowered peas the offspring are all yellow. But whether the parent with the yellow flowers supplies the egg or the pollen makes not the slightest difference. Yellow still dominates. Male and female gametes are entirely equivalent. In the 1860s this result was far from obvious. A great deal of folklore still prevailed; and one ancient folktale said that *all* the inheritance came from the sperm, and the egg's job was merely to nourish. The absolute genetic equivalence of the two parents was in absolute contrast to such ancient conceits and was well worth emphasising.

For more than 100 years Mendel's idea of absolute sexual equivalence was genetic dogma. But dogmas in science are not sacrosanct and in the 1980s it became clear that in some cases it *does* make a difference whether a particular gene is inherited from the mother or the father. The *same* gene inherited from a different parent may have a different effect in the offspring. Many examples are now known, some of which are of direct medical relevance. Huntington's chorea, for example, is caused by a single dominant gene that typically remains unexpressed until late in life, and then comes on line with crippling damage to the nervous system. But the disease is more severe, and strikes earlier in life, if the gene is inherited through the male line rather than through the female line. Indeed if the gene is passed from father to son through the generations the disease becomes more and more severe, striking earlier and earlier; but if it goes from father to daughter then the daughter's children will have milder doses, later in life.

This phenomenon, whereby the parent puts his or her own stamp on the gene, according to their sex, is called 'genomic imprinting'. Its existence is surprising: it is not what a century of genetics studies had led biologists to expect. It came to light largely through experiments in nuclear transfer (many by Azim Surani then at the Institute of Animal Physiology at Babraham, Cambridge) in which mammalian zygotes were given two female pronuclei, or two male pronuclei. Such zygotes can develop as far as the blastocyst stage but if they are then implanted into a uterus they soon fail. The embryos with an all-paternal genome develop a fine placenta, but no proper embryo; while those with an all-maternal genome begin by making a good-looking embryo, but only a poor placenta. For this reason parthenogenesis in mammals – development of a whole new animal from an unfertilised egg – really does seem 'biologically impossible' in mammals, even though it is common enough in other animals, including many other vertebrates. In the absence of divine intervention, virgin birth for mammals is not an option. Both a male and a female genome must be present. (However, parthenogenesis might be achieved in the future if biologists learn how to convert a maternally imprinted genome into a paternally imprinted genome, or vice versa. No one is even close to this but there is no present reason to doubt that it is possible.)

Genomic imprinting has given rise to an important line of evolutionary theory, largely at the hands of David Haig in the United States. In general, biologists now perceive that although the different genes of the genome cooperate with each other – of course they must – they also compete. Selfish genes really are selfish. Paternally imprinted genes seem to induce mechanisms and behaviours that are good for – well, paternally imprinted genes; while maternally imprinted genes seem to induce mechanisms and behaviours that are good for female-imprinted genes. (Though note in passing that females contain genes that are paternally imprinted, as well as maternally imprinted, and males contain maternally imprinted genes as well as paternally imprinted genes. It can be quite confusing!)

But in the early 1980s genomic imprinting was still a new idea, and so was mammalian cloning, and it seemed as if the former might interfere with the latter. After all, genomic imprinting is a form of genetic programming; and if genes have to be reprogrammed when nuclei from older cells are transferred into oocytes or zygotes, why shouldn't they have to be reimprinted as well? Thus McGrath and Solter say in their 1984 paper, 'any transferred nucleus would have to be able to reprogramme male- and female-specific genomic activity', and 'It is very unlikely that such precise reprogramming . . . can be achieved in a nuclear transfer experiment.' Indeed the complete text of their sentence in which they said that mammalian cloning cannot be done reads: '*Differential activity of maternal and paternal genomes,* [as well as] the results presented here, suggest that the cloning of mammals, by simple nuclear transfer, is biologically impossible.'

In fact, as Dolly and many other cloned mammals now demonstrate, genomic imprinting is not always an insuperable problem; Dolly at least is alive and well and her first lamb, Bonnie, seems eminently healthy and normal. The imprinting that has already been imposed on the transferred DNA seems to escape reprogramming – at least in those embryos that developed to term. Imprinting and cell differentiation both involve the closing down of genes but the mechanisms are evidently distinct. Among other things, imprinting is known to take place during gamete formation, whereas differentiation takes place in the early embryo. We should not be cavalier about this, however. Genomic imprinting may be responsible for some cloning failures – perhaps many; and we have not looked at enough cloned animals for long enough to be certain that it will have no effects later in life.

Fortunately, not everyone believed that mammalian cloning by nuclear transfer was 'impossible'. One who emphatically did not was Steen Willadsen.

Steen Willadsen

Steen Willadsen is a Danish veterinarian who worked as a research scientist first in Britain and later in the United States. He is also a star – a visionary, it has been said. Fittingly, as a vet, he has worked mainly with farm livestock and, for reasons that we have already glimpsed, sheep and cattle seemed to be better candidates for cloning than mice. Sheep are easier to handle than cattle, and cheaper.

Way back in 1979 Willadsen, then at the ARC's Institute of Animal Physiology at Cambridge, described how he produced five pairs of monozygotic twins of sheep – that is, five sheep clones of two individuals each – and by so doing developed techniques that have been employed in large animal cloning ever since. As he explained in *Nature* on 25 January 1979, he did *not* produce these early sets of twins by nuclear transfer. He simply separated the blastomeres of two-cell embryos. In more detail: he held each two-celled embryo at the tip of a pipette by suction, and used a glass needle to tear open the zona pellucida along the groove that runs along the divide between the two cells. When the tear extended along about three-quarters of the way round, he sucked out the blastomeres with another pipette, one at a time.

Now came an important technical innovation. He aimed to develop entire embryos from each of the blastomeres – which was well known to be possible in principle since identical twins do occur naturally in sheep, as well as in humans. But mammalian embryos normally spend their first days in the oviduct. However, if the zona pellucida is damaged, then at least until they reach the morula stage, they are destroyed by the mother's immune system. Here he had blastomeres whose zona had been torn almost in half. So how could they develop further? One solution would be to grow them *in vitro* until they were ready to be implanted into the uterus of a ewe. But at that time young sheep and cattle embryos could not be grown satisfactorily *in vitro* – and indeed this is still difficult.

Willadsen's solution (albeit based, as he acknowledges, on a technique that had been developed for mice in the 1960s) was first to put each blastomere into a zona that had already been evacuated; and then to encase the resulting 'embryo' in a protective plug of agar (agar being the jelly obtained from seaweed that biologists traditionally use to hold media in which to culture all kinds of cells). In fact, Willadsen made a series of agar plugs each of which contained identical twin pairs of blastomeres – each being one-half of a two-cell embryo and each protected by a new zona. He trimmed each plug to measure 0.15 by 0.5–1.00 mm, and then immersed these small plugs in agar again, to make bigger plugs: 0.7 by 2.0–2.5 mm.

He then transferred thirty-one of these plugs – each containing its twin pair of blastomeres – into the oviducts of ewes that were in the first or second day of their oestrus cycle: the stage at which, if they had conceived, they would have two-cell embryos in their oviducts. He put between one and four plugs into each oviduct, tying the oviduct on either side of them to make it easier to recover them later. Thus these ewes acted as incubators. After leaving them there for 3½ to 4½ days he opened the ewes up again, and flushed out the plugs. Embryos in twenty of the thirty-one plugs had developed normally; in all, he recovered thirty-five embryos that by now were late morulae or blastocysts; and these included sixteen pairs in which both embryos had developed.

The agar coats were still intact in all the embryos that he was able to recover, and he prised the coats off with hypodermic needles. Now he transferred each of the sixteen pairs of twinned embryos to the oviducts of sixteen ewes that were in the fifth, sixth, or seventh day of their oestrus cycle. Each ewe received one set of the monozygotic twins.

Eleven of the sixteen ewes became pregnant. One aborted on day hundred of pregnancy. But the other ten animals all gave birth – five of them just to singleton lambs, and five to pairs of monozygotic twins.

Conceptually, of course, cloning by embryo splitting is much simpler than cloning by nuclear transfer. But the true importance of this early work by Willadsen lies in his technique. Scientists accumulate techniques like magpies, and we at Roslin have picked up a lot from Willadsen. Thus Megan and Morag, Dolly and Polly and the rest were also made by protecting the reconstructed embryos in agar, and then transferring them to the ligated oviducts of temporary recipient ewes, before transfer to the surrogate mother that gives birth to them. At Roslin, as we will see, Bill Ritchie does most of the embryo reconstruction; and, he says, recovering the embryos from the agar after their spell in the temporary recipient is one of the trickiest parts of the operation (although the rest of it is certainly tricky enough).

Willadsen continued to innovate throughout the 1980s. He produced bigger clones by embryo splitting, once deriving a set of quin lambs from an eight-cell embryo. Eight is probably as high as you can go by this means, partly because the genome starts to express at this point and partly because each divided blastomere is expected to act effectively as if it were a zygote, but a blastomere from an eight-cell embryo has only one-eighth of the original cytoplasm. Willadsen also made a series of chimeras: combining cells from different embryos and even mixing cells from different species – sheep with goat, for example – to create mixed blastocysts that sometimes developed into whole and even mature animals.* One of his goat-plus-sheep chimeras, known colloquially as a 'geep', featured on the cover of *Nature*. Roughly at this time I (Ian) was considering how to make chimeras with embryo stem (ES) cells. Most important of all, however, is that Willadsen was the first biologist unequivocally to clone mammals by nuclear transfer.

Willadsen: the first sheep cloned by nuclear transfer

Willadsen was the first biologist to carry out nuclear transfer in sheep, and in *Nature* on March 1985 – a decade almost to the day before Keith and Ian's formal announcement of Megan and Morag – he described the first successful use of the technique to

---

*Chimeras are not to be confused with hybrids. A hybrid is produced when two creatures (they might be animals or plants) of different species, or from different varieties within a species, are crossed. In the resulting offspring, each and every body cell contains copies of the genes from both parents. Hybrids are often sexually sterile, although they may not be, because the two different genomes cannot cooperate properly during meiosis. Thus horses and donkeys may mate to produce mules or hinnies; but mules and hinnies are sterile because although the horse and donkey genomes are compatible enough to produce body cells they are not compatible enough to provide gametes. (And note too, in passing, that mules – with big bodies and small heads – are produced by crossing horse stallions with donkey mares; while hinnies – with big heads and small bodies – are produced by donkey sires and horse dams. This looks like a good example of genomic imprinting.)

Chimeras, in contrast, are made by mixing cells or other body parts from two different individuals. The contributing individuals may be of different species – as in Willadsen's geep; or the donors may be of the same species. Chimeras of spectacularly diverse 'parentage' feature in many a mythology, though not in reality: centaurs, griffons, Indian gods with the heads of elephants. At a more homely level, they feature in every suburban garden: a grafted rose is a chimera, though both contributors are of the same species.

A chimera made by mixing different cells is said to be 'mosaic'. But, whereas each and every cell in a hybrid contains copies of the genomes of both parents, each cell in a geep, say, is either pure sheep or pure goat. Cells that are introduced into young morulae to make a chimera sometimes become commandeered as germ-line cells. Since they are pure-bred cells, they are perfectly capable of meiosis. Thus a geep, say, might produce pure goat eggs or sperm, or pure sheep eggs or sperm, or a mixture of both. Here then, among other things, lies a technique that could be of use in conserving rare species: introduce cells from one of its embryos into embryos from a common, related species; produce chimeras; and, from those chimeras, obtain pure eggs or sperm of the original rare species. Willadsen was and is concerned about conservation.

clone sheep. Illmensee's reports of mice could not be repeated, so Willadsen's cloned lambs can reasonably be seen as the first mammals of any kind ever to be cloned beyond any doubt by nuclear transfer. Megan and Morag were an advance on Willadsen's first cloned sheep in three outstanding ways: the donor cells were taken from much older embryos; they were multiplied in culture before transfer; and – most significantly of all – the Roslin scientists paid particular attention to the cell cycles of the donor and recipient cells, which Willadsen did not. But Willadsen's cloned lamb was seminal nevertheless, in concept and in technique. Such pivotal work merits a brief description.

First Willadsen injected unmated Welsh Mountain × Cheviot ewes with human chorionic gonadotrophin (hCG) hormone to induce them to ovulate and, 30 to 33 hours latter, he collected unfertilised eggs from them. These became the cytoplasts. Then he mated other Welsh Mountain × Cheviot ewes with a Suffolk ram and from these he collected eight-cell embryos on day three, and sixteen-cell embryos on day four. These provided the karyoplasts. Thus the genetic provenance of the embryos that provided the karyoplasts was different from that of the oocytes that were to provide the cytoplasts. The former were Welsh Mountain × Cheviot × Suffolk, while the latter were simply Welsh Mountain × Cheviot.

He then needed to remove the nuclei from the cytoplasts but, since sheep oocytes are opaque, he could not simply reach in with a pipette and suck them out. He cut through the zona pellucida just above the polar body, using a glass needle, and then placed the cytoplasts in a medium containing cytochalasin B to 'relax' the cytoskeleton. After an hour or so he used a pipette with a tip diameter of around 30 microns – about a third of the size of the oocyte itself – to suck out about half of the cytoplasm that lay immediately beneath the polar body. About three times in every four the half that he removed contained the metaphase chromosomes; and the half left behind was enucleated.

To make the karyoplasts he isolated single blastomeres from the eight- or sixteen-cell embryos and, with the aid of syringes, placed them in contact *either* with an enucleated cytoplast *or* with one that still contained its original nucleus.

He then employed one of two techniques to fuse the karyoplasts with the cytoplasts. Either he used Sendai virus, as by then was conventional; or he employed an 'electrofusion' technique that had been developed in the 1970s. The two cells are simply placed side by side between two electrodes and subjected to an electric shock. The electric current breaks down the outer membranes of the cells, temporarily, just as Sendai virus does; and – as with the virus – the membranes fuse as they re-form, like the skins of merging soap-bubbles. The trick is to expose the cells first to alternating current, which brings them into alignment, and by polarising them draws them together, then to give them a few brief blasts of direct current, which effects the fusion. Electrofusion proved far superior: only half of the embryos treated with virus actually fused, and this took up to four hours, while ninety per cent of the electrified embryos fused within an hour. So here is another very significant Willadsen innovation. He did not invent electrofusion, but he was the first to employ it for reconstructing embryos. Willadsen spotted the electrofusion apparatus in an advert in *Nature*. For us at Roslin, electrofusion (slightly modified from Willadsen's original technique) is now standard. Whatever the method of fusion, Willadsen placed the reconstructed embryos in cytochalasin B to facilitate their merger.

Thus he had a batch of reconstructed embryos, each made with about half the cytoplasm of an oocyte, some containing nuclei only from the donor blastomeres and some containing a blastomere nucleus plus the original egg nucleus; and all were enclosed in damaged zonas. (Of course he needed extra zonas to accommodate them all, which he got from other oocytes.) So now he could employ the trick he had developed in the 1970s: enclose each reconstructed embryo in a plug of protective agar and put it into the oviduct of a 'temporary recipient' sheep, the duct tied either side to form an incubation chamber.

Four-and-a-half to five-and-a-half days later he recovered the embryos from the ewe oviducts. Seventy-six of these had been made from blastomeres taken from eight-cell embryos and, of these, thirty-two (forty-two per cent) had now formed blastulae. Twenty-nine of the recovered embryos had been made from blastomeres taken from sixteen-cell embryos and, of these, fourteen had now formed blastulae: around fifty per cent. So far the success rate was encouraging.

What was truly exciting, though, was that the blastocysts – whether they had been made from eight-cell or from sixteen-cell blastomeres – were normal: they looked and behaved just as they would if they had developed from normal zygotes. Perhaps the nuclei that came from the eight-cell and sixteen-cell embryos had simply retained their totipotency: genomic activation does not occur in sheep until the eight- or sixteen-cell stage. But perhaps, too, the nuclei had already changed somewhat by the time they were transferred – and, if so, then they had clearly been reprogrammed by the oocyte cytoplasm in which they had been placed.

Now came the *pièce de résistance*. Willadsen took four of the blastocysts that had been reconstructed from eight-cell nuclei, and introduced them into the oviducts of surrogate mothers. Three of them developed into full-term lambs. Two of them were identical twins: clones of each other. This was in the 1983–4 breeding season. By 1986, when his report was published in *Nature*, he had cloned several more – some from sixteen-cell blastomeres.

Willadsen concluded: 'It seems reasonable to suggest that a firm basis has been established for further experiments involving nuclear transplantation in large domestic species.' He had freed cloning research from the tyranny of the unsatisfactory mouse, and brought it into the realms of agriculture, with all its opportunities and resources.

As a final flourish, Willadsen said that he had already taken blastomeres from *reconstructed* blastocysts, and used them to make a second generation of blastocysts. This is 'serial cloning', which – at least in theory though almost certainly not in practice – would make it possible to produce clones with an effectively limitless number of individuals. The American researchers, who sought to clone elite cattle, found such serial cloning particularly attractive.

Willadsen played an even more direct role in the research at

Roslin, which we will discuss shortly. But we should look briefly at parallel attempts across the Atlantic – to clone cattle.

The American approach: cattle

Sheep have proved excellent subjects for cloning but there is no point in cloning them just for the sake of it. In general, animals that are as cheap as sheep would not justify the huge investment needed to replicate them. We have remained interested in sheep partly so as to build on the technologies of genetic transformation, and also in part as a model for cattle. For the economics of cattle are quite different, and on paper they certainly justify cloning. Accordingly, a group in the United States that began under Professor Neal First at the University of Wisconsin and since has spread to many other institutions – involving Randall Prather, Frank Barnes, Jim Robl and others – focused specifically on the cloning of elite cattle by nuclear transfer right through the 1980s and into the 90s. Four biotechnology companies grew up to develop the technology, of which the biggest were W. R. Grace and Company and Granada.

Cattle form one of the biggest industries in the world. Their performance depends on their genes – the environment too, of course; but they cannot respond to, say, a richer diet unless they are genetically equipped to do so – and there is plenty of genetic improvement still to be made. The top milkers in even modern herds may yield twice the average amount of milk – so the present task is not to raise the top level but to improve the average. But genetic improvement takes an enormous amount of time because cattle breed so slowly. A cow is pregnant for nine months and can produce only one calf a year. In the case of dairy cattle, too, it takes several years to see if a cow is good enough to be worth breeding from – and even more years of 'progeny testing' (that is, assessing the performance of the daughters) to see whether a bull is a useful stud. The decades can flash past. Artificial insemination, first applied commercially to cattle in the 1940s, was the first technological attempt to accelerate genetic improvement, and is now the norm in the dairy herds of rich countries. Bovine *in vitro* fertilisation was developed in the 1990s, and is still not very reliable. Embryo transfer has been practised since the 1970s – generally employing embryos produced *in vivo*, by superovulation. Cloning cattle of proven performance seems the natural extension of these techniques. On the face of things, its impact worldwide could be tremendous.

So the American emphasis on cattle cloning has made perfect sense, and has enjoyed some impressive successes. In 1987 First and his colleagues announced that they had cloned cattle from eight- and sixteen-cell embryos; in 1989 they produced quins from an eight-cell embryo; and in 1994 they produced live cattle off-spring from ICM cells. But, despite the expertise and the huge investment, the cattle-cloning programme did not establish itself as a significant technology, and certainly it has not made the worldwide impact that was once envisaged. The most obvious problem, the *coup de grâce*, was a quite unexpected side-effect. Although 'normal' cattle embryos produce normal pregnancies when transferred to surrogate mothers, embryos reconstructed by nuclear transfer are slow to be born (pregnancy is prolonged) and, commonly, are thirty per cent larger than normal, or even more. Birth then becomes difficult and painful and, of course, the economics of the whole operation goes to pot. As Keith says, 'It makes no sense to add 100 dollars to the value of the calf and then pay 200 dollars for a caesarian.' Largely because of large fetus syndrome the commercial companies dropped out, the last to go being Genmark in Salt Lake City. Practical biology is indeed full of serendipities – it would be lost without them. But it is riddled with unforeseeable pitfalls, too.

In sheep, the extent to which cloning produces over-large fetuses is difficult to quantify. The size of a lamb is influenced both by its own genes and by the genetic background and physiology of its mother – and when a lamb is carried by a surrogate mother of a different breed it is difficult to say what size such a lamb *ought* to be, and so to gauge the extent of any abnormal enlargement. Thus Dolly was a Finn-Dorset lamb carried in the womb of a Scottish Blackface surrogate mother. Both breeds are about the same size although, as it happens, the Scottish Blackface lambs born at Roslin have tended to be slightly bigger at birth than Finn-Dorset lambs. At birth Dolly was large by Finn-Dorset standards as experienced at Roslin, but not by Scottish Blackface standards. Large fetus syndrome clearly is an issue in sheep and its causes, and how to prevent it, are being studied at Roslin by Lorraine Young.

Colin Tudge:

All in all, the early 1980s were the turning point on the road to Dolly. In 1980 it was not clear that mammalian cloning by nuclear transfer was possible at all; then, under Illmensee and Hoppe, it seemed easy; then, after Solter and McGrath, it seemed impossible; but by 1985 Steen Willadsen had done it, not with mice as Illmensee had attempted but with sheep, which for this particular purpose seemed far easier; and soon afterwards, Neal First and his colleagues had done the same in cattle. Keith, while all this was going on, moved from a doctoral project at the Marie Curie Institute in Surrey to Sussex University where he began a second DPhil, which he took to completion. Ian plugged away at the lab that was to become Roslin, trying to improve the technology of animal transgenesis. Both, for different reasons, kept weather eyes on the cloning work. Ian saw its potential for his own endeavours – 'and in general I'm comfortable with the idea that research should have a practical goal'.

In 1986, again by a slightly circuitous route, Ian began his own cloning programme at Roslin.

CHAPTER 7

Cloning Comes to Roslin

Ian Wilmut:

In 1982 – ten years into my professional career as a scientist, and after nine years at Roslin – I felt that things were going well. I was trying to find out why so many embryos die *in utero*: an important topic, both in medicine and in agriculture, which is still hotly pursued in other laboratories. My work was pure physiology and senior colleagues had criticised me for not incorporating enough genetics – this was the fashion, then – but external assessors, of the kind who every now and again report on government research laboratories, rated the work very highly.

Then Roger Land, newly appointed as director of ABRO (as Roslin then was) told me to bring that work to an end. His own interests lay mainly in the genetics of reproduction; he had worked on this at Edinburgh under Douglas Falconer and Ann McLaren after graduating from my alma mater, Nottingham. At that time, as we have seen, ARC was keen to cut support for agricultural research in general; and Roger wanted to introduce more molecular biology into the labs. He told me I had to work on genetic transformation: either that, or go. It was that blunt. My job was to provide the zygotes (produced in ewes by superovulation) and then to add the DNA to them. I was not at all happy. I was carving my own niche as a research embryologist and I resented the way he brought my research to an abrupt end. I did not want to spend my life adding genes to zygotes, and trying to produce live animals from them, essentially as an adjunct to somebody else's project. Besides, injecting DNA into one-cell embryos is technically exacting because zygotes are only the same size as the oocytes from which they derive, which means they are about a tenth of a millimetre across; and although this is giant by the standards of most cells it is still mighty small and my hand tremor prevents me from doing any serious manipulations on such a scale. In short, this was just not the kind of work I enjoy. With hindsight, of course, the change of direction was fortunate both for me and for the institute – but at the time it did not seem that way. But I agreed with Roger from the outset that once I had helped to establish a protocol for adding DNA to the young embryos, and transferring them into surrogate mothers, I would look for more efficient ways to do the job; and this at least should be creative.

Even so, I thought long and hard about leaving; but I talked it over at length with Vivienne, and at last we decided to stay. We enjoy the Scottish borders. We both play active parts in the life of the village. I belonged to the local farmers' club and had taken up curling. Later I became president of both clubs. We love the countryside, and walking our dogs in the hills. I was doing quite a lot of jogging and even ran two marathons – the Glasgow and the Edinburgh – with a best time of 4½ hours: just a little more than twice the world record. Our children were doing well at secondary school (the Scottish education system is much envied) and Vivienne was involved in the Presbyterian Church (and later became an elder). Vivienne and I are English northerners but socially we had become good Lowland Scots.

But also, I began to envisage how to advance the project that I had been thrust into, while also satisfying my own desire for original research in developmental biology. Answer: don't just add DNA to one-cell embryos. Add it to plates of cultured cells, and then make embryos from the cells that had taken up the DNA most effectively. In other words, as the 1980s wore on I began to see that the future of genetic engineering in animals lay through cloning.

In the early 1980s, two conceptually different areas of work seemed particularly pertinent. The first was the science and technology of embryo stem (ES) cells, which biologists were beginning to use to effect genetic transformations in mice; the second were the several forays, in various laboratories, into cloning by nuclear transfer. Much later, in January 1987 – it was in a Dublin pub; I know the time and place exactly, as I will describe later – I realised that we could put the two technologies together and so use nuclear transfer as a way of carrying out genetic transformations in animals. This is what we have done in Roslin – culminating in Polly, born in 1997, who incorporates aspects of both lines of work; although in the end, as we will see, we more or less abandoned the search for ES cells and achieved success by a route that I had not initially conceived, following a quite different line of thought that developed with Keith.

But I am running ahead. I should describe how I became involved with the technologies of ES cells and of cloning by nuclear transfer in the 1980s, long before Keith arrived on the scene. I will take one at a time.

The search for ES cells

In the early 1980s Martin Evans and Matt Kaufman at Cambridge showed that it was possible to culture cells from the middle region of young mouse embryos – inner cell mass (ICM) cells – in ways that retained their totipotency. This is far harder than might appear. It is difficult enough to maintain cells in culture at all; they do not necessarily thrive. Assuming they do survive, they tend to change in culture. Basically this is because, as we have seen, the genome is responsive to its surroundings: depending on the conditions, some genes will be switched on, others switched off. Thus the cells adapt to the prevailing conditions. So to maintain totipotency in culture it is in essence necessary to imitate the kind of conditions found in the young embryo – which doubtless provides a flood of chemical signals in critical proportions that are certainly not understood in detail. Evans and Kaufman cultured the mouse ICM cells on 'feeder' cells – support cells that we now know provide specific factors that regulate the growth and stability of the cells.

Although these ICM cells did alter somewhat in appearance as they were cultured, they did retain the generalised, rounded look of embryo cells. More to the point, when these cultured cells were put back into the inner cell mass of another embryo, they *behaved* exactly like normal ICM cells. As the young embryo develops, the cells of the inner cell mass are able to develop into *any* of the newly differentiating tissues: muscle, liver, germ cells (eggs or sperm) – what you will. The cultured ICM cells of mice were able to do this too. Even after they had been cultured, these cells could also develop into muscle, or liver, or germ cells, when they were put back into an embryo. Hence their name: 'embryo stem cells'. Stem cells in general are cells that retain totipotency – or at least, more usually, a high measure of pluripotency. ES cells are simply stem cells that come from embryos.

This discovery of ES cells raised tremendous possibilities – which the animal genetic engineers were quick to realise and exploit. For, although it is difficult to add genes to a zygote and produce a truly transgenic embryo, it is astonishingly easy to add genes to cultured cells, at least in a rough and ready way: in fact, if you simply pour DNA over the top, some of the cells will incorporate it into their genomes. If you add DNA to a zygote you get only one chance; if it works you get a transgenic embryo, and if it doesn't you don't. But if you culture the cells first, then you can produce tens of thousands of them – so you have tens of thousands of chances (or millions, if you like!) to produce the required genetic changes in at least one of the cells. Furthermore, when you work with zygotes, adding DNA is *all* you can do, at least with present techniques. But mere addition of DNA is only one part of the genetic engineering repertoire: we would also like to be able to remove or 'knock out' particular genes, or to take out DNA and change it and put it back. If you have tens of thousands of cells to play with, and if they can be kept in culture for many weeks – as indeed they can – then in principle it is possible to carry out *any* required genetic manipulation.

The problem, of course, is to turn the transgenic cell in culture into a whole animal. If the cultured cells happen to be ES cells, then this can be done – albeit by an indirect and somewhat chancy route. Once you have made the required genetic changes in a cultured ES cell (and identified the particular cells in which the novel transgene is working well) you merely have to reintroduce those cells back into the inner cell mass of another embryo. Then, with luck, *some* of the reintroduced, transformed cells will differentiate to form germ cells in that embryo. Thus the embryo will grow to form an animal whose eggs or sperm contain the required transgene. If this animal is then allowed to breed by normal sexual means you can produce an entire dynasty of animals that all contain that transgene. As already described in chapter 6, animals that have had cells introduced from other animals are called 'chimeras'; and chimeras produced by incorporating cells from another embryo grow up with two kinds of cells in their tissues and are then called 'mosaics'. In the early 1980s, genetic engineers regarded such chimeric mice as a godsend: they contained ES cells that had been genetically transformed in culture, and they produced offspring that contained the required transgene from the outset. We wanted to make transgenic sheep – so why not do as the mouse biologists were doing? Why not produce ES cells from sheep, manipulate their genes in culture, and then make sheep chimeras who could sire or give birth to truly transgenic offspring?

There seemed to be just two obvious snags. First, it would be very time consuming to make transgenic sheep by the chimera route that works so well in mice. First the donor cells had to be cultured and genetically transformed, which could take weeks or months. Then the chimeric embryos would be created, and put into a surrogate mother, and gestation would take five more months. But the chimeric lambs that would result from the pregnancy are mosaics: only some of the cells contain the transgene. The truly transgenic animals are produced in the next generation – from the genetically transformed eggs or sperm of the chimeras. So then we would have to wait for the chimeric animals to mature – another 18 months; and then produce lambs of their own – another five months. These second-generation lambs would then be fully transgenic. But if the aim is to derive valuable products from their milk – as with Tracy – we would then have to wait another 18 months for those lambs to grow up. Mice run through two entire generations in a few months; but in sheep two life cycles take several years. The decades would soon flash past and, in these harsh commercial days, time matters.

Even so, the ES route to transgenesis offered clear advantages, and was obviously worth pursuing.

But there was a second snag, even before we got round to tackling the problems of logistics. ES cells were known from mice but no one in the 1980s had produced cells from any other kind of animal that would behave like an ES cell, and retain totipotency in culture. Of course, all mammalian embryos have ICM cells; but the ICM cells of most animals rapidly differentiate when they are cultured, and lose their totipotency. When reintroduced into the inner cell mass of other embryos, they may multiply but they do *not* differentiate to form a varied range of tissues. Cells have been isolated from the embryos of other animals that have some of the properties of ES cells, and some of those other kinds of cells have been called ES cells; but none, from any other species, has ever shown itself to be fully totipotent when reintroduced into the inner cell mass of another embryo. Indeed, 'true' ES cells with this outstanding quality had been isolated from only two particular strains of laboratory mouse.

But in the early 1980s there was no reason to suppose that ES cells could not be produced from other strains of mice – or from any other species; and since ES cells offered such obvious theoretical advantages in genetic engineering I spent a great deal of time in the mid 1980s and into the 90s, working with various other scientists, trying to produce ES cells from sheep.

It seemed as if the problem was simply to find the right conditions for culture, which would enable ICM cells to multiply without differentiating. Cell culture after all is a craft as much as it is a science; to a large extent improvements are made just by adding things and taking things away, and seeing what results. The behaviour of the cells in culture can be modified by changing the conditions, just as a gardener can influence the behaviour of his plants. I had no reason to doubt that, if we tried long enough, we should be able to produce ES cells from sheep. There was just no reason to suppose that this was impossible.

Cell culture properly began in 1885 when Wilhelm Roux (he who sought to demonstrate Weismann's theory of tissue differentiation – as described in chapter 4) showed that he could keep embryo cells of chicks alive in a saline solution. Then about 20 years later R. G. Harrison made what can perhaps be seen as the crucial step into modernity: he cultured cells from amphibian spinal cord on a clot of lymph, and showed that in this cultured state they produced nerve fibres, or axons. Harrison was seeking to demonstrate that axons do indeed emanate from single cells, and are not compounded from many different cells, which was the alternative hypothesis. But in practice he established two general principles of cell culture. First, the cells of animals are reluctant to grow in culture unless they are in contact with some solid surface – and in fact it has become standard practice to grow animal cells in plastic dishes, which provide a sympathetic surface. In this respect, animal cells are quite unlike those of bacteria, which grow happily in suspension.

Secondly, Harrison was beginning to show that animal cells rely heavily on messages from surrounding cells: infusions of various 'factors' which prompt their growth and direct their differentiation. He grew his nerve cells not simply in 'saline' (solutions of salts) but in contact with living tissue (or extracts from tissues), which provided essential growth factors – although no one at that time had any idea what those factors actually were.

Nowadays there are vast manuals of techniques for culturing cells of many different kinds for many different purposes, and the necessary media with ingredients of uncanny purity can be bought by mail order. All in all the procedures are far more precise and 'scientific' than in Harrison's day. Yet cell culture is still a craft, or indeed an art, like gardening or cooking. Recipes for culturing particular cells for particular purposes tend to be very complicated and precise (with exact proportions of particular metal ions, precisely controlled pH and temperature and so on) but they also include ingredients that have a cabalistic quality, and have not and possibly cannot be analysed exhaustively, yet are known to provide essential components commonly summarised as 'growth factors'. Fetal calf serum is one such ingredient; and there are many others. In addition, as already mentioned, cells are often grown in the company of other, 'helper' cells which provide various 'factor X's' that remain unanalysed in detail. In truth, if you have a culture medium that contains twenty or so different ingredients, and the ratio of each to each is important, but some of them are unknown (like the particular contributions of helper cells) then there probably is too little time in the Universe to analyse exhaustively what each ingredient is contributing, and what difference it makes to change any one component ever so slightly. Cell culture therefore tends to be a conservative craft: if a system works, then stick to it. In the history of cell culture serendipity has often played its part: refrigerator doors left ajar or (as when penicillin was discovered), windows left open allowing microbial invaders to enter. All in all it is a hugely intricate and arcane pursuit.

But a few large generalisations emerge. First, cells change when they are in culture. Animal cells commonly take on a kind of generalised skin-like form, known as *fibroblast*. Dolly's contemporaries, Taffy and Tweed, were made from cultured fibroblasts derived from fetuses. Cultured embryo cells generally lose their totipotency; at least, if they are simply transferred from culture into oocytes or zygotes they will not produce embryos that survive for very long – unless they are suitably prepared first, as with Megan and Morag. In general, cultured cells undergo only a fixed number of divisions before they stop growing altogether: different cell lines, from different species, have their own characteristic life span. Some cell lines, however, achieve what appears to be immortality. The most famous cultured cells of all are the HeLa cells, which were first established from a human cancer in the 1950s and are now to be found – vastly altered from their original state – in laboratories throughout the world.

Cultures are established by placing a few cells on to a suitable medium and then allowing them to multiply. As we have said, animal cells, unlike bacterial cells, will not generally grow unless in contact with an underlying substrate. The cultures generally remain just one cell thick, and they stop multiplying when the sheet of cells reaches the limits of the container. Then they have to be lifted from the medium with the help of enzymes, which attack the proteinous 'glue' with which cells stick themselves to their substrate, and 'plated out' in to fresh media – just as a gardener might repot a plant, only the process is somewhat more elaborate. The interval between each replating is termed a 'passage' (pronounced, as in the French fashion, 'pass-*arje*'). The cells undergo several cell cycles (meaning that they divide several times) in the course of each passage; each cycle typically lasts a day or several days, so each passage commonly lasts a week or more. So cells that have undergone a dozen or so passages – as was true of the cells that produced Megan and Morag – will have been in culture for some months. If biologists want to transform the cells genetically (as in the case of Polly, but not of Dolly or Megan and Morag) while they are in culture, then they have plenty of time both to manipulate the genes as required and to monitor the results. Once a culture is established it can be frozen and kept indefinitely. When it is thawed, the biologists can resume work where they left off (provided everything is done with consummate skill).

One bonus of my work in the early 1980s was the pleasure of working with some extremely accomplished scientists. They included Matthew Kaufman, at Edinburgh, who we have seen was one of the first to develop ES cells from mice; and Martin Hooper, also from Edinburgh, who was one of the first biologists to use mouse ES cells to introduce genetic changes into animals so as to study genetic diseases and the role of gene products in health and sickness. A third was Alan Handyside from the Medical Research Council at Carshalton in Surrey, part of a team who have developed methods to screen early human embryos and thus remove those that carry specific genetic mutations.

Through the mid 1980s Alan, Martin, Matthew and I collaborated to produce cultures of sheep cells that we hoped would behave like ES cells of mice; my role was to recover the embryos from the donor animals and be able to assess any ES cells that were isolated. We summarised much of this effort in 1987 in a paper entitled 'Towards the isolation of embryonal stem cell lines from the sheep' (in *Roux's Archives of Developmental Biology*). In that particular study we began by culturing ICM cells from sheep embryos, and we included a refinement already developed by Martin Hooper and Austin Smith who found they could inhibit differentiation in cultured ICM cells of mice by adding liver cells from rats.

We first treated Welsh Mountain ewes with hormones to make

them ovulate (or indeed to superovulate), then mated them, and then, six or seven days later, flushed the young embryos – blastocysts – from their oviducts. (This is the basic procedure for obtaining blastocysts; we will describe it in more detail in chapter 9.) In this case we wanted to get at the ICM cells; and they of course lie in the middle of the blastocyst, surrounded by trophectoderm (TE) cells. So these TE cells first have to be stripped away, which we do using an immunological technique that other scientists had developed elsewhere. First you prepare antibodies against sheep tissues by injecting sheep cells into cows, and you can then extract the anti-sheep antibodies from the serum of the cow's blood. If you now add these antibodies to the blastocysts they attack the TE cells first, since these are on the outside, and so expose the ICM cells within. As soon as the ICM cells are exposed you wash off the antibodies, and away you go (if you get the timing right!).

So we set up nine different cultures of ICM cells. The cells thrived – but they did not retain the rounded, ES-like form of mouse ES cells. Instead, in the typical manner of cultured animal cells, they assumed the form of fibroblasts. There were a few rays of hope, though: some of the cultured ICM cells did retain an ES-like form through the first passage. But as soon as the culture had filled the dish, and had to be plated on, the cells became fibroblast like.

This is about as far as we got. Indeed this is roughly as far as anyone has got. At Roslin, Jim McWhir continued a quest for ES cells in livestock that he began with Martin Evans in Cambridge. David Wells completed a very successful PhD project in which he found ways of increasing the efficiency with which ES cells could be isolated from mouse embryos – which could be very important when we begin to use human ES-like cells for human cell therapy, as discussed later. David then returned to his native and much-loved New Zealand (and also to his girlfriend, who later became his wife) where he subsequently established one of the most successful projects for cloning cattle.

But neither we, nor anybody else, has ever managed to produce true ES cells from species other than one or two strains of mice – even though some other cell lines from various species, with some

ES-like properties, have sometimes been called 'ES cells'. The search continues and, as we will see in chapter 9, Jim McWhir argued that some of the sheep we produced at the same time as Megan and Morag were made from cells that had ES-like qualities; although he called them '*TNT cells*' – meaning 'totipotent for nuclear transfer' – rather than ES cells. But Keith maintained that Jim's cells could not retain totipotency through more than a few passages and, even if they did, that this was not necessary. For a time, the argument was fierce. But that's science. Sometimes sharply contrasting hypotheses can seem equally plausible. Sometimes the one that seems less likely turns out to be correct. There are bound to be disagreements.

But we will come to this. Suffice to say that, all in all, ES cells from sheep and cattle remain elusive. We have not yet produced ES cells from any species other than mice, and Keith is inclined to argue that the whole attempt has largely been a waste of time. I don't agree. Negative results are useful too. We have learnt a great deal from our attempts to produce ES cells, and I have no doubt that if we could produce them they would still have considerable advantages. It would also be very instructive to know why it is possible to produce ES cells from some mice, and not (so far) from other mice or other species.

The second line of research that clearly seemed relevant were the various forays, in Europe and the US, into cloning by nuclear transfer. I became directly involved with this work through my contacts with two other scientists – both of them vets, as it happens: the Brazilian Lawrence Smith, who came to work at the Roslin; and Steen Willadsen, who I had known from my days at Cambridge.

Cloning, Ian and Lawrence Smith

Lawrence Smith came from Brazil, though from a British family. He was already qualified as a vet but his heart was in research and he went first to Edinburgh to complete an MSc in animal breeding. There he attended lectures by Ron Hunter, who was then at the School of Agriculture in Edinburgh. Ron was a former colleague from my days in Cambridge and he recommended Lawrence to me. He arrived at Roslin (still called ABRO) in 1985 to read for a PhD in reproductive physiology; and I was to be his supervisor, guiding him through the doctoral degree. PhD students come in all shapes and sizes. Some, unfortunately, have simply chosen the wrong line of work. Others are obviously excellent from the start (it can be very hard to keep up with them) and these in general are of two kinds. Some come into your office every week with a new idea, and the problem is to get them to finish one thing before they start the next; others, once started, just like to be left alone. Lawrence was of the latter kind: very bright; very competent, and wanted to do his own thing.

When he first came, I suggested that he might look at factors in the oviduct that influence egg development: a neat and important topic that could produce a worthwhile thesis. But Lawrence had his own ideas. As we have seen, the hot topic of 1985 was nuclear transfer: not simply with a view to cloning, but also as a route into many fundamental questions. This, he said, was where his interest lay.

To begin new lines of work, PhD students have first to produce a research proposal – and Lawrence's was most impressive. In general, he wanted simply to understand more clearly how embryos develop; and in particular to unravel the dialogue that passes between nucleus and cytoplasm. He proposed several lines of attack. Neal First had recently proposed that the development of embryos could be divided into conceptual stages, which he called 'domains'. Nuclei could be successfully transferred between embryos that were in the same developmental domain, but not between embryos in different domains. Thus a nucleus from an eight-cell embryo might continue normal development if transferred into a four-cell embryo, but a nucleus transferred from an ICM cell would fail if transferred into a four-cell embryo, having already passed into a new developmental 'domain'. McGrath and Solter's experiments outlined in chapter 6 illustrate some of the idea. Lawrence sought to see how the domain principle might apply in mice. By such means, he said in his proposal, he sought to explore 'the value of the oocyte in reprogramming nuclei' for,

Aerial view of Roslin Institute, Edinburgh

Professor Grahame Bulfield, director of Roslin Institute

Diagram of the nuclear transfer procedure that produced Dolly

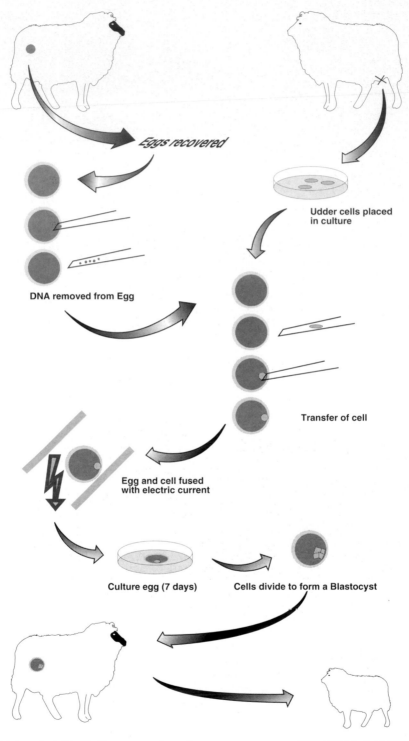

Eggs recovered

DNA removed from Egg

Udder cells placed
in culture

Transfer of cell

Egg and cell fused
with electric current

Culture egg (7 days)   Cells divide to form a Blastocyst

Egg transferred to Blackface surrogate mother   Giving birth to lamb (Dolly)

Bill Ritchie alongside the microscope and micromanipulation equipment he used during the experiments described in this book

(Left) Harry Bowran, senior farm research assistant; and (right) Ray Ansell and Jim McWhir, who grew the cells used to produce Megan and Morag

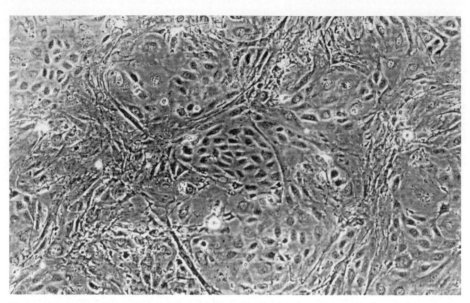

Embryo derived cells used to produce Megan and Morag as seen through the microscope

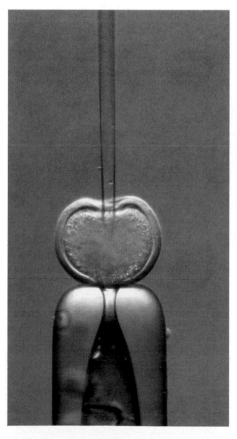

Cloning in progress. The larger pipette holds the egg in place while the smaller one removes the genetic information – the process of enucleation

Dolly, Megan and Morag with John Bracken, Marjorie Ritchie and Douglas McGavin

Dolly's wool was removed by world champion sheep shearer Geordie Bayne in aid of the Cystic Fibrosis Trust. The Leishman sisters have the illness and were intrigued that a sheep might be able to help them

The two rams, Taffy and Tweed, the first offspring to be produced from fetal fibroblast cells

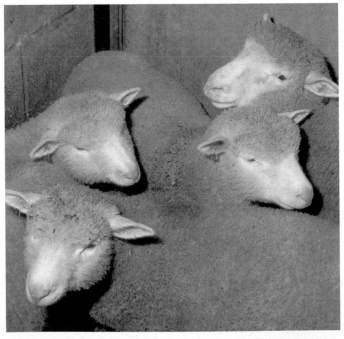

Polly and sisters, each carrying an additional gene that produces a clotting factor in their milk

Ian Wilmut with Polly

Keith Campbell

Jaki Young (left), Ian Wilmut's personal secretary who also ensures the smooth running of the team, and Lorraine Young, who has a research project to understand and eliminate the causes of failure in cloning and other assisted reproduction techniques

Dolly with her first lamb, Bonnie

Dr Harry Griffin, assistant director (science) of Roslin Institute,
with Dolly and her three lambs born in 1999

he said, 'Though there is not currently any published evidence from mammals it is probable that oocyte cytoplasm has the ability of reprogramming nuclei of later embryonic cells' ... [and] ... 'an initial trial will be set to investigate this idea and to verify the extent of nuclear differentiation which the oocyte can totally reprogramme.' He also intended 'to search for tissues whose nuclei are totipotent or closer to totipotency than the current sources of embryonic cells'.

This was excellent: ambitious, important – and, as it happened, in line with my own growing interests. There was a political problem, however, because ABRO at the time was still linked to the Babraham laboratory in Cambridge, and nuclear transfer seemed to be in their remit rather than in ours. But I sought permission from Roger Land and, after some negotiating, we were given the go-ahead. So it was that nuclear transfer – the basis of the technique that eventually produced Megan and Morag, Dolly, and Polly – came to Roslin. Lawrence Smith was clearly a key player.

But Lawrence's task was not easy. For one thing, he chose to carry out the essential cell fusion with Sendai virus – but Sendai virus is wretched. You can't buy it off the shelf, and it is difficult to grow. The turning point in the cloning work came through another stroke of serendipity, in January 1987, in a Dublin pub. There, I got to learn of crucial new research by Steen Willadsen, then working in Canada.

Dublin 1987: and the meeting with Steen Willadsen

John King was appointed director of ABRO when I joined in 1973, and I admired him greatly. Roger Land succeeded him in 1982, and remained until Grahame Bulfield, who is still director of Roslin, took over in 1988. Roger was a distinguished reproductive geneticist and had many fine qualities, but he and I did not always see eye to eye. My first boss had been Chris Polge at Cambridge, that affable man who is so sharp and clever but also so kind. Chris's attitude was relaxed; he appointed the people he liked, and let them get on with it. His approach suited me and it certainly suited Steen

Willadsen, who is one of biology's great individualists, and who succeeded to my fellowship at Chris Polge's laboratory. I have adopted the same kind of style myself at Roslin – I like to let people get on with things. Perhaps unconsciously I am copying Chris. But Roger Land was rather different and wished to be involved in everything. If the task is merely technical, then that's fine: you need to get everyone working to the same routine. But creative scientists need to be left alone for at least some of the time, and guided with a light touch.

Despite our differences, however, I will always be in great debt to Roger for making the opportunity that ultimately led to the birth of Dolly. As a consultant and speaker he earned occasional fees which, by the rules of those days, directors of institutes were not allowed to keep. So Roger put these earnings into a travel fund, to enable ABRO scientists to attend conferences. I liked to attend the International Embryo Transfer Society's meetings (I was the creator of Frostie, after all!) but the society normally met in the US, which was somewhat expensive. In January 1987, however, the society decided to meet in Dublin, just a short hop over the Irish Sea, and Roger invited me to accompany him. We had written a paper together. Roger was to deliver it, while I chaired the session.

As in all such meetings, the best things happen on the sidelines. One evening in the pub I found myself with a group of scientists that included Geoff Mahon, whom I knew from my Cambridge days; and indeed he was still there, as a physiologist. Geoff in turn was a close confidant of Steen Willadsen. Geoff commented that Steen, now working in Canada, had just cloned a calf by transferring a nucleus from an ICM cell.

The notion that science proceeds by a series of 'eureka moments' – dramatic insights that seem to come from nowhere – is not necessarily the case. Often ideas just seem to creep up on you. But when Geoff described Steen's work I felt pennies dropping and bells clanging. Let me remind you of the state of play at that time. Karl Illmensee had reported a few years earlier that he had produced mice by nuclear transfer from ICM cells but his work had been disputed – and apparently firmly refuted by McGrath and Solter. Neal First and his colleagues in the US were producing calves by nuclear transfer – but they were taking nuclei from very young embryos, well before the blastocyst stage. The belief – indeed the dogma – of the time was that viable mammalian embryos could not be produced from cells that were taken from embryos that had passed the first few cell divisions; and the reason for this, which seemed both logical and essentially proven, was that in the very young embryo the genes are switched off, but after the first few cell divisions the genome is switched on. Once the genome is switched on it starts to run through the programme that directs the life of the animal; and once the programme starts running, differentiation begins, and totipotency is quickly lost. The dogma of the day said, in fact, that viable embryos *could* not be made from ICM cells, since ICM cells were well past the stage of genome activation. But Willadsen, it seemed, had done it. Apparently there was something wrong with the dogma.

So now put this observation together with the idea that I had been following. I had been trying to make ES cells from sheep so as to make sheep chimeras, just as had been done in mice; and ES cells are made from ICM cells. But now Willadsen had apparently made viable embryos from ICM cells of sheep *without* making chimeras, with all the uncertainty and loss of time that this entails. Instead, he had simply made new embryos by nuclear transfer. If he could make sheep embryos by nuclear transfer, from ICM cells taken straight from a blastocyst, why shouldn't I make embryos by nuclear transfer from ES cells, which were cultured ICM cells? In practice, after all, a blastocyst yields only a few ICM cells that could produce new embryos; but in culture, we could produce as many as we wanted. Furthermore – which was the crucial point for my work at ABRO – it would be possible, once the ICM cells were in culture, to make genetic changes at leisure. So I saw the future: culture ICM cells from sheep (or cattle) and so produce ES cells; effect genetic transformation; and then use nuclear transfer to make new embryos from the transformed cells. Thus we could make precise genetic changes in animals – and then, by cloning, we could make groups of animals that were genetically identical.

I sat next to Roger Land on the plane home from Dublin and I told him that I just had to visit Steen and find out exactly what he had done and how. I am sure that Roger agreed because he could see that this was a way both to use my experience with embryos and to advance his own ambition – to make a significant contribution to animal breeding through molecular biology.

A few weeks later I had to visit an Australian government laboratory outside Sydney (one of the CSIRO labs) to offer advice on the production of transgenic animals. In fact by the time I arrived they had solved the problem and I was left with time to read, think, and visit other labs in Adelaide and Melbourne. This was all very pleasant. Much more to the point, though, was that I could visit Steen on the way home, at his lab in Calgary.

Steen is a gentleman of science: he loves the subject and he loves talking about it. His wife is a scientist too – Carol Feehilly, who was a student in Chris Polge's lab when Steen was there. She now works in human IVF. They are both extroverts, hospitable and fun to be with. Steen carried their young child with him as we walked and talked in the scrubland that lies near to his laboratory and to his fine, large house. Yes, he said, he had indeed made calves by nuclear transfer from ICM cells. Later he showed me technical details: I watched him coat the newly constructed embryos in agar for protection, then place them into the oviducts of the ewes that acted as their temporary recipients. As we will see in chapter 9, when we talk about Megan and Morag, these became our own standard techniques.

However, although the target began to seem clear after my discussions with Steen – culture ICM cells to make ES cells, and then use them to clone embryos by nuclear transfer – a lot of groundwork still needed to be done. On the one hand, it still was not proving easy, or indeed possible, to make ES cells from sheep. On the other, Steen's success in cloning from ICM cells seemed to raise a whole new raft of questions which I felt had to be addressed if we were to turn bright ideas into reality.

In particular, it remained clear that as cells multiply – whether in an embryo, or in culture – they do tend to differentiate, and to lose totipotency as they do so. My own approach to science is, I would say, logical and methodical; and it seemed necessary to ask when, precisely, a lineage of cells does lose totipotency, both during embryogenesis and in culture. When we knew when, we could ask why totipotency was lost: what were the underlying mechanisms. From the late 1980s, then – working at first with Lawrence Smith – I set out to answer these questions.

Roslin's first clones – and a vital change of direction

I have already commented that Lawrence introduced cloning by nuclear transfer to Roslin; in the late 1980s he worked both with sheep and with mice. His methods were definitely state of the art. Sendai virus was commonly used for cell fusion in mice but it was less effective in other species and, in his sheep experiments, Lawrence abandoned Sendai virus in favour of electrofusion. He used cytochalasin B to soften the cytoplasm, and aid enucleation. And so on. Bill Ritchie is now the master of these techniques, but he learnt them originally from Lawrence.

Lawrence and I produced Roslin's (then ABRO's) first cloned lambs in 1989, as recorded in *Biology of Reproduction*. We produced four live, cloned lambs by nuclear transfer. Three of them came from blastomeres from sixteen-cell embryos; and one – emulating Steen Willadsen – came from an ICM cell. This work was crucial for several reasons. It established our institute as a serious centre of cloning and it showed, once more, that embryos *can* be produced from ICM cells – well after transcription has begun, and the genome is activated. We all take it for granted, now, that this is possible; but in the late 1980s this was still a hot subject for debate.

Lawrence completed his PhD thesis in that same year and was duly awarded his doctorate. Next year, 1990, he took up a post in Canada, where he remains at the faculty of veterinary science at the University of Montreal. Beyond doubt, he was a key figure in the events that led up to Megan and Morag and to Dolly.

But now comes the final twist to the story: the crucial change of emphasis. At the end of the 1980s I still felt, along with virtually all biologists in the field, that embryos could not be made successfully by nuclear transfer unless the nuclei retained totipotency. This seemed to be the clear lesson of the previous 100 years. True, John Gurdon had produced frogs from cells that had long since gone past the point of totipotency; but frogs are not mammals and it seemed to me – and still does – that we are much more likely to obtain embryos from totipotent cells.

So in the late 1980s I felt (and still do) that it was vital to explore how and when totipotency is lost as cells replicate, in embryos and in culture; and until well into the 1990s this seemed to be the vital question. So with Jim McWhir (and later with Keith) I embarked on experiments to clone embryos from cells from older and older embryos, and from cultured cells that had been through more and more passages, seeking to pin down the vital cut-off point when totipotency was finally lost.

Yet Lawrence's work at ABRO in the late 1980s already began to hint that the age of the cells – the number of divisions they had undergone, and the extent of differentiation – was not the only factor we needed to attend to. Something else was important, too; and that something else was the cell cycle.

In those days Lawrence and I had very little idea indeed of what the cell cycle actually is. The phenomenon is in the domain of cell biology, and although you might suppose that developmental biologists would also be cell biologists, this really is not the case: science has become more specialised than that. Besides, even within cell biology, ideas on the cell cycle were still in the formative stages.

Lawrence was beginning to show, however, that the success of nuclear transfer was influenced by the state of play within the karyoplast nucleus. It clearly mattered whether the donor cells had just divided, or were preparing to divide again, or were somewhere in between. It was becoming clear, too, that the stage of the cycle of the karyoplast should ideally correspond to that of the receiving cytoplast. Thus Lawrence records in his thesis that 'transplantations using nuclei and cytoplasm from different cell cycles produced more blastocysts when the recipient cytoplasm came from later stages of the first or second cell cycle' and 'it may be necessary for DNA replication to be completed before mitosis and cytokinesis [the final separation of the cells].' He was getting there. But the cell cycle was not the prime subject of his PhD project and this was as far as he could get in the time available.

But it became clear to me that, if we were truly to fulfil our goals, we needed to know more about the behaviour of cells. Lawrence left in 1990 and on 14 February 1991 we placed an ad in *Nature* for a 'post-doctoral higher scientific officer' to study 'gene expression in early embryos'. My memory is that the ad asked specifically for expertise in the cell cycle; but when I fished it out for the purposes of this book, I was somewhat surprised to find that it does not. It simply says that 'Mechanisms regulating gene expression and chromatin structure in the embryo and that improve the efficiency of nuclear transfer procedures in cattle will be studied. The person appointed will be required to use molecular biology techniques and/or embryo micromanipulation'. The ad concludes somewhat mysteriously: 'Relevant skills would be an advantage' – as opposed, presumably, to irrelevant skills, or no skill at all. The project was to be funded initially for three years and was sponsored by various bodies and involved collaboration with the Scottish Agricultural College and the University of Reading. As I commented earlier, the administration and financing of modern research can be at least as complicated as the science.

I cannot recall how many people responded to the ad but I do know that we finished up with a shortlist of four – all impressive in different ways. Among them was a DPhil from Sussex called Keith Campbell. His hair was longer in those days but then my beard was bushier so I suppose we were about equally hirsute. One of my interviewing techniques – I learnt it from Roger Land – is to ask the candidate to describe some piece of work, whether or not it is related to the matter in hand, just to see if he or she can put it into a clear conceptual context. The long-haired man from Sussex undoubtedly could (although as Keith tells it, 'I ranted on for about an hour and a half and they all fell asleep'). My daughter, Naomi, still at school, was at that time spending time at Roslin (by that time it was called IAPGR) for 'work experience'; and she joined Keith and me for lunch. She, too, was impressed by the clarity of his ideas.

My notes on the candidates record on the plus side that Keith had 'previous experience with embryos, studied cell division, familiar with molecular biology techniques' but under 'Cons' I listed, 'very few publications, despite five years since completion of DPhil. Unusual background.' But I was impressed. In particular, his DPhil thesis seemed to be just what we needed: 'Aspects of cell cycle regulation in yeast and *Xenopus*' (*Xenopus* being the same much favoured laboratory frog that John Gurdon had used for his pioneer cloning experiments). I wrote to Grahame Bulfield (who took over as director in 1988), 'I have studied all of the applications [and] this has helped to clear my mind as to what I am trying to achieve.'

So we invited Keith to join us as a postdoctoral research scientist, and so he began on 6 May 1991, a couple of weeks before his thirty-seventh birthday.

Colin Tudge:

Thus began one of the most fruitful working partnerships in modern biology. This, in an age that has seen Crick and Watson, Jacob and Monod, and Kendrew and Perutz (who worked out the three-dimensional structure of myoglobin), is no mean claim. Ian had the grand plan – to make sheep from cultured cells by nuclear transfer – and Keith brought to the endeavour a crucial understanding of the cell cycle. To appreciate Megan and Morag, Dolly, and Polly (and to see why biologists have now been able to clone mice from adult cells, even though outstanding scientists failed to make them from embryo cells in the early 1980s) we should take a chapter to look at the cell cycle in some detail.

CHAPTER 8

Keith and the Cell Cycle

Colin Tudge:

The cell cycle is the complete agenda of the cell: cell division, duplication of DNA, and then (in most cases) more cell division. Its significance has crept into people's consciousness over the past few decades. John Gurdon in the 1960s makes no mention of the cell cycle; he is concerned primarily with the provenance of the donor cells – whether they are taken from young embryos or tadpoles or adult frogs – rather than their individual state of play. But many of the scientists involved in mammalian cloning in the 1980s did think about it. McGrath and Solter acknowledged its possible significance in their paper of 1984 and Lawrence Smith at Roslin was clearly beginning to perceive its possible significance in his PhD thesis of 1989.

For the cloning biologists of the 1980s, however, the cell cycle had to remain just an uncomfortable awareness. At that time no biologist understood in detail how the cell cycle worked, or how to control it – and still less why the phase of the cell cycle would affect the outcome of nuclear transfer. There was a growing body of literature on the cycle, but there was no manual of cell cycle science and technology to take off the shelf. Neither is it the kind of subject that a biologist can simply study on the side. The cell cycle is complicated and esoteric. It requires full-time dedication. In short: the science and technology of cloning needed input from a different branch of biology – the study of cell cycles. Not every cloning biologist realised that this was the case and, even for those who did, the necessary information simply did not exist.

But biologists in labs worldwide were on the case – and Keith was among them. He brought his expertise to Roslin in 1991; developed his ideas in the context of the cloning project; and showed, eventually, that in cloning by nuclear transfer the cell cycle is not just an aside, a detail requiring a weather eye, but one of the keys – perhaps *the* key, to the whole operation. This is an extraordinary reversal. We should ask how Keith came to be involved.

Keith Campbell:

My own route into science has been a little unconventional. Most scientists take a few A levels at school, then a BSc (or BA) at university, which typically takes three years, then perhaps an MSc (usually one year), and then a PhD. In some countries graduate students begin their PhDs in their late twenties, and may take five years or more; but aspiring scientists in Britain commonly begin a PhD in their early twenties, and are 'doctored' within three years. So British scientists typically begin their first professional job as a 'postdoc' in their late twenties, either in a university, or a government institute (like Roslin) or in commerce.

I by contrast contrived to leave my grammar school in the Midlands without A levels. I went there on a scholarship, but I never really warmed to the atmosphere of the place. I left to take an Ordinary National Certificate (ONC) – a vocational exam – in Medical Laboratory Sciences, which I got at the age of 19 in 1973; and two years later I took the Higher National Certificate (HNC). In other words I trained initially not to be a research scientist, but a technician: the kind of person who helps the researchers to do their work. But it was not intellectually satisfying. I wanted to do my own research. So I got a place at Queen Elizabeth College, London to read Microbiology and the day I got my HNC – which qualified me fully as a technician – I resigned my job (I was then working in a hospital). My technical training stood me in good stead at Queen Elizabeth, however. I found, for example, that some other students might leave a culture of cells lying about over the weekend and then be surprised, when they came back on Monday, to find it festooned with mould. I just could not do that sort of thing.

So I qualified in 1978 at age 24 with a 2.1 (not bad, not brilliant) and looked around for something interesting to do; I first took a job in a pathology laboratory in Yemen. Again I had some good times; again I gained experience in handling cells (and was given far more responsibility than I was supposed to have) but I knew it was not the kind of thing of which careers are made. I returned to England a year later and then spent another year (1979–80) helping to control Dutch Elm disease in Sussex – this being a fungus disease, carried by beetles, that within a few years in the early 1980s virtually wiped out what had been one of England's commonest and most characteristic trees. My main job was to identify the most threatened trees, and seek to isolate them, in the hope that they would not become infected. Again it was fun and interesting; but again, in the end, unsatisfying.

So in 1980 I began my first PhD – at the Marie Curie Institute in Carshalton, Surrey. The institute included a number of groups who studied aspects of human cancer, using tissue culture and various 'animal models'. I was a research assistant in the Tissue Culture and Cytogenetics Group. I assisted by day – one of my long-term projects was to investigate the effects of the birth-control pill on human cells – and studied for my PhD in the evenings and weekends. It was hard but it could have worked out well – except that my boss, Dr Nutan Bishun, became very ill (with heart trouble) and was away for much of the time, and finally had to retire. Then the institute's director retired and the new director in the spirit of a new broom undertook what I thought was a well-overdue reorganisation. Many of the scientists left, and although I stayed on for another six months I found myself working on my PhD on my own, which is not the way to do these things. So the Marie Curie Institute generously gave me a grant to continue my studies elsewhere – the first, and I believe the only, Marie Curie Research Fellowship.

But although my career at Marie Curie petered out in my three years there (1980 to 83) I developed the insights and desires that were finally to shape my career. The first was the control of cell division. Most body cells, most of the time, proliferate and differentiate in the ways appropriate to the organ of which they are a part, and then they stop proliferating. Cancer cells seem to recognise no such controls. They continue to multiply, and form tumours. Why?

Secondly, I had been taught – it was in effect the dogma – that differentiation is irreversible. Once a body cell is committed to become liver, or skin, or whatever, then that was an end of it until the cell died. But some tumours contain many kinds of cells: liver, finger nail, hair and so on. Yet it seems likely that tumours generally arise from a single cell – and from a cell, furthermore, that has already differentiated. In such tumours, then, the original cell has dedifferentiated, multiplied, and then the daughter cells have differentiated again. Differentiation, in short, was not so irreversible as dogma had it.

Thus I realised that it was time to change the direction of my career. I would study the control of cell growth, proliferation and differentiation: or, more specifically, I would study the cell cycle. PhD students need grant money – Marie Curie had given me that – and they need a supervisor: a scientist in some established institution to act as guide and mentor. The leading exponent on the control of cell division at that time was Dr Paul Nurse at the University of Sussex, who had been working with the fission yeast *Schizosaccharomyces pombe* (*S. pombe*): a very useful organism in which to isolate and study the genes that are associated with the control of cell division. But Paul was planning to move to the Imperial Cancer Research Fund (ICRF) in London and did not want to take on PhD students until he had established his new lab. He suggested instead that I should talk to Chris Ford at Sussex about a project that the two of them were planning, in the replication of DNA in the South African clawed frog, *Xenopus laevis*. Chris was also interested in the process that, much later, proved to be of key significance in the cloning work at Roslin – the maturation of oocytes: the way in which the young oocytes, locked in prophase in the first stage of meiosis, proceed to the stage at which they are ready to be released – metaphase of meiosis II (or MII). The immediate task was to identify the gene or genes involved in oocyte maturation.

So I took this project on and in July 1983 I began my second PhD, this time at Sussex (where PhDs are called DPhils). The oocytes of frogs – the whole early development – proved fascinating. I read the literature, and became acquainted with the work of John Gurdon. In 1984 Karl Illmensee visited Sussex to lecture on cloning in mice; Gurdon and Illmensee between them kindled my interest in this field. I have fond memories of Sussex – playing pool and, much more to the point, the excellent discussions on science. As we will see shortly, the cell cycle is complicated – that's the way life is! – and in the early 1980s it was barely understood at all; so the complexities manifested as loose ends. It would take too long to try to explain how all the things we studied at that time fit into, and contributed to, the modern picture, indeed it would require a separate book. As will shortly be described, however, a key agent in the control of the cell cycle is MPF (what the letters mean will soon become apparent). MPF, as is now clear, is a universal: it is effectively the same in all organisms – although in those days, this was the kind of thing we were still finding out. Thus I sought at one point to isolate MPF from the fission yeast, S. pombe, to see if it would induce the oocytes of Xenopus, the clawed frog, to mature. In the end I could not isolate MPF from the fission yeast, but I did get it out of brewer's yeast Saccharomyces cerevisiae and, yes, it does work in the frog. Mitosis lasts longer in brewer's yeast than it does in fission yeast, so MPF is present in the cells for longer, which is probably why it is easier to obtain from brewer's yeast. Data from these experiments also showed that MPF has two components – as again we will see shortly.

Chris Ford's interest in DNA replication led to an active research programme in which I became involved with Chris Hutchinson, who joined soon after me as a postdoc; and we linked up in turn with Ron Laskey and Julian Blow at Cambridge. Chris Hutchinson and Julian in a series of experiments showed the importance of the nuclear membrane – the nuclear envelope (or NE) – in controlling the replication of DNA; and when we added the MPF that I had isolated we could induce mitosis. I had many fine discussions, too, with Ian Kill, who was (and is) working on the senescence of cells; and, again, our experiments and discussions later proved relevant to nuclear transfer, the technology that underpins modern cloning. Again, all will become clear. All in all they were heady days – the way I always felt science ought to be: excellent problems, keenly pursued. Our group did not elucidate the cell cycle all by ourselves – this would be very rare in modern science! – but we were certainly part of the endeavour; and, as will become apparent, the problems we worked on at that time became key insights in the research that led to Dolly.

I completed my PhD studies in 1986 and submitted my thesis the following year – its sombre title hardly reflecting the intellectual excitement that lay behind it: 'Aspects of cell cycle regulation in yeast and *Xenopus*'. Then I got my first job as a professional scientist – 'postdoctoral research fellow' – in the zoology department of Edinburgh University. Edinburgh was the birthplace of much of the work on cell cycle control. I joined Peter Fantes's group, to work on the control of the cell cycle in the fission yeast again, *S. pombe*. A key figure at Edinburgh was Murdoch Mitchison, with whom I spent many hours discussing and hypothesising on MPF. At one time, Paul Nurse, Kim Nasmyth, and Peter Fantes all worked in Mitchison's lab.

But postdoc research fellowships run only for a limited time; and in 1989 I had to leave to take another postdoctoral fellowship at the University of Dundee. I was sorry to leave – except that I was again able to work with Chris Hutchinson, who was now a lecturer at Dundee. With Chris, my research moved back to frogs. We sought to isolate various extracts from the cytoplasm, with which to control the behaviour of the nucleus, and of the DNA within it: exploring in general the dialogue between cytoplasm and nucleus, and in particular, the control of the cell cycle. In early frog embryos cell division is extraordinary. The complete genome replicates in twenty minutes – whereas it takes hours in somatic cells. The whole process is emphatically led by the cytoplasm – which continues to divide whether DNA replication is completed or not. So replication has to be rapid. If you place somatic nuclei in the cytoplasm of early embryos they ignore the signals at first, and do not replicate their DNA. But if you leave them for four to five hours they change, structurally, and become more like the nuclei of embryos again – and then they do replicate their DNA, and will indeed complete replication in twenty minutes. Such insights may seem peripheral to the cloning of sheep (and at that time I had no particular thoughts of cloning sheep) but we will see their pertinence. All this work helps to show the innate and unexpected flexibility of the nucleus and of the genome it contains. Send the right signals to the genome, and it will adjust. Somatic nuclei that would not normally replicate can be induced to do so; and (we might reasonably extrapolate) genomes that are already differentiated might be reprogrammed.

Then in January 1991 I saw Roslin's ad in *Nature*. It was for a proper salaried job and of course was interesting; I had been thinking about cloning since my Sussex days, and here was the chance to apply my ideas on the cell cycle. Besides, my partner, Angela, and I had just had our first daughter so I needed something steadier; and Roslin was only 16 miles from our home, while Dundee was 80 miles away to the northeast. I applied and was taken on.

It may seem a big leap from yeasts and frogs to sheep and cattle, but the point is that it is not. If you work on enough cells – from different species, both normal and cancerous – you see the similarities. In particular, the features that control the whole agenda of the cell, the cell cycle, are to a large extent universal. Knowledge of the cell cycle had largely been acquired from yeasts and frogs but once we applied that understanding to the problems of nuclear transfer in sheep we truly began to see how cloning could be achieved. Now, almost a decade later, it seems that if you pay proper attention to the cell cycle – primarily of the donor karyoplast but also of the recipient cytoplast – then the factors that hitherto had been thought to be so obviously vital – the age and differentiation of the donor cell – seem to be secondary. Whether the donor cell comes from a young embryo, a fetus, or an adult animal; whether it is cultured before transfer or is not cultured; and, if it is cultured, whether it goes through one passage or a dozen or more, may not matter – or not, at least, certainly not as much as everyone had anticipated. If you adjust the cell cycles of donor karyoplast and recipient cytoplast, you can produce viable, reconstructed embryos from differentiated cells – and perhaps, with better understanding and technique, from any kind of cell. Megan and Morag (made from cultured and differentiating embryo cells), Taffy and Tweed (made from cultured fetal fibroblasts), and Dolly (made from cultured cells derived from an adult mammary gland cell) provide the demonstration.

So now we should look at the present understanding of the cell cycle: bearing in mind that even by the early 1990s, when we first began the series of experiments that led to Megan and Morag and then to Dolly, the picture was far hazier than it is now.

The cell cycle

Rudolf Carl Virchow proposed in the mid 19th century that all cells are the daughters of other cells, so that each cell's cycle begins with the division of its parent cell. Most cells grow after they first arise, and when they have grown sufficiently, divide again. The division of the chromosomes within a typical body cell is called mitosis, the details of which were described in chapter 5; and this is followed by division of the cytoplasm, generally known as cleavage. The phase between divisions, when the genes produce proteins and the cell grows, is called interphase. Thus a cell in a growing tissue, or in culture, goes through mitosis, interphase (including growth), mitosis, interphase, and so on. The general pattern has been evident since the end of the 19th century, and this is about as much as most biologists who are not cell specialists know about the cell cycle even today.

But in 1953 – the same year that Crick and Watson described the three-dimensional structure of DNA – Alma Howard and S. R. Pelc at Hammersmith Hospital, London, showed that there is a lot more to the cell cycle than this. They worked with growing tissue (*meristem*) of onion root; but what applies in this context to an onion root cell applies equally to the cells of sheep mammary gland or human skin or what you will. Howard and Pelc proposed that interphase – which most biologists tended to think of as one single event – can logically be divided into three phases that are quite distinct, and have quite different functions.

What, after all, can be seen to happen in mitosis? Well, the chromosomes condense, which is when they first become visible under the microscope. But when they first appear, they are seen to be *doubled* structures, each one consisting of two chromatids, joined at the centromere. 'Chromatid' may sound like a diminutive, but in fact each chromatid functions as a complete chromosome; so the chromosomes really are doubled. But during mitosis the chromatids in each chromosome separate; and in each new daughter cell that emerges after mitosis the chromosomes are only single structures. In other words, each erstwhile 'chromatid', now that it has separated from its twin, is afforded the proper status of a chromosome (which, functionally, it was all along).

So note a few points. First, every biologist knows that most body cells of most animals are diploid; they contain two sets of chromosomes. Only in a few special cases (such as the drones – males – of bees) are the body cells of animals haploid. However, when the chromosomes first appear before mitosis they are *doubled* structures, so that *each* chromosome in fact contains *two* functional copies of the DNA. Thus, in effect, before mitosis a normal body cell of an animal contains *four* functional sets of chromosomes. In other words, at the build-up to mitosis, it is *tetraploid* ('*tetra*' being Greek for 'four'). Many plants have tetraploid cells as a matter of course, including wild species such as the common saltmarsh grass *Spartina townsendii*, and many domestic crops, such as the European domestic potato *Solanum tuberosum*. Tetraploid plants are sometimes more vigorous than diploid ones (although there must be some disadvantages too, or tetraploids would be more common than they are). But tetraploidy in animals, and especially in mammals, is generally thought to be rare. However, *every* normal diploid animal cell that divides by mitosis is functionally tetraploid in the period before mitosis, because each of its chromosomes has doubled. Plants like the European domestic potato whose default position is tetraploid are octoploid in the build-up to mitosis.

When do the single-structured chromosomes that are bequeathed to each new daughter cell become the doubled structures that are seen in the build-up to mitosis? And what is involved in the doubling?

Well, first of all, we tend to have a biased view of chromosomes. They are visible under the microscope only when they are condensed – tightly coiled and supercoiled. They take on this condensed form only when they are preparing for mitosis (or meiosis); and when they are preparing for mitosis they have already doubled, so they appear as doubled structures. But in its pristine state each chromosome is a single structure. Indeed, each chromosome consists of a single macromolecule of DNA (although of course, as Crick and Watson revealed, each single macromolecule consists of a *double* helix).

During interphase the macromolecules are decondensed: the DNA molecule is still coiled – in the sense that it consists of a double helix – but the long, helical molecule is spread out throughout the nucleus. It is not supercoiled, and bunched up on itself, as it is in the build-up to mitosis; and in this decondensed, spread-out state it is not visible with the light microscope. It is in this spread-out, uncluttered state, during interphase, that the DNA is free to do its main job: producing RNA, which then makes protein.

But also, some time around the middle of interphase, the DNA replicates itself. Instead of producing RNA, it produces more DNA. In practice, the two helices in each double helix are prised apart with the aid of enzymes, and then more enzymes build an exact complement to each separated helix. Actually, the complementary helices are built even while the two original ones are still separating, so no DNA helix is ever really alone. A 'zone of separation' moves along the DNA double helices, forming new complementary DNA strands as it goes. Thus each double helix becomes two double helices, with the two new doublets remaining joined at the centromere. Of course, Howard and Pelc could not envisage all this as described here because the molecular details became clear only after Crick and Watson had published their model. But Howard and Pelc did monitor the duplication of

DNA by labelling one of its component modules, thymidine, with a radioactive isotope.

The phase in which the DNA is duplicated, Howard and Pelc called the synthesis (S) phase – since new DNA is synthesised.* So now we perceive two crucial phases: mitosis and synthesis. When the cell enters mitosis it is functionally tetraploid, and comes out diploid; and when it enters synthesis it is diploid, and comes out tetraploid.

Howard and Pelc then noted, however, that logically (and functionally) there is a phase after mitosis and before synthesis, and this they called 'gap 1' (or G1): and a phase after synthesis and before the next mitosis, which they called 'gap 2' (or G2). So the complete cycle is mitosis, gap 1, synthesis, gap 2, mitosis – or, for brevity, M, G1, S and G2, followed by another M, and so on. Interphase thus consists of G1, S and G2. In G1 the typical body cell of an animal is diploid; in G2 it is tetraploid; and in S it is progressing from diploidy to tetraploidy.

Now for a few vital twists. During the S phase, *the DNA replicates once, and only once*. If it fails to complete replication, or begins a second round of replication before entering G2, then the daughter cells that are finally produced after the ensuing mitosis will have too few or too many chromosomes, or its chromosomes will be deformed. A cell with the wrong number of chromosomes is said to be *aneuploid*; and, although some plants can withstand some degree of aneuploidy, in animals the loss or gain of chromosomes, or chromosome damage, is usually lethal or at least severely compromising. After mitosis it is in general essential that each daughter cell finishes up with the right number of chromosomes. If this is to be achieved, then mitosis must not begin until S phase has finished; and, equally, S must know when to call a halt.

---

* 'Synthesis phase' is an unfortunate term. After all, the DNA is said to 'synthesise' protein; but 'synthesis' in Howard and Pelc's terminology refers to the DNA's synthesis of more DNA. It would have been better to call this phase the 'replication phase'. But they did not. Biology is unfortunately packed with somewhat misleading terms – names that typically are applied to particular entities or functions before their full significance has been worked out. Anyway, the synthesis or S phase is when the DNA copies itself.

One final elaboration, vital to our story: the cell cycle can be interrupted at various stages – both by artificial means, in the laboratory, and naturally. In particular, cells in G1 sometimes go into a kind of shut-down or *quiescent* state known as G0 (which some people pronounce 'G-zero' and others call 'G-nought'). In practice, G0 probably describes cells in various states: some cells that are presumed to be in G0 may simply be moribund but others are apparently switched off and buttoned down in a very orderly way, like a ship cocooned in dock or an animal in hibernation. Cells seem to enter G0 when they are under stress; cultured cells can be put into G0 by reducing their level of growth factors, for example. In any case, G0 cells show very little activity, which perhaps is why cell biologists have paid them less attention than they probably should have done. In cancer, for example, it is the frantic *growth* of cells that seems important, not their periods of quietude. Yet as we will see, cells in G0 have played a key role in the production of the cloned sheep at Roslin, from Megan and Morag to Dolly and Polly. It also turns out that G0 isn't *just* a buttoned-down, 'hibernating' state: indeed the term 'quiescent' may be misleading. G0 seems to be an essential stage that all cells go through in the course of differentiating; and this seems to be why it has proved so vital in the creation of Megan and Morag, Dolly, and Polly. But we will come to this.

This, then, is how the cell cycle operates in typical body cells, whether in animals or plants or fungi such as yeasts, both within the living creature and in culture. But, although the principles are the same in all cells, in specialist cells with specialist functions the details may differ: notably, during the formation and early development of the embryo. The details clearly matter a great deal to biologists interested in cloning, and we should look at them.

The cell cycle at conception and in young embryos

A new life begins with the fusion of sperm and egg. As we have seen in chapter 5, the sperm is haploid at the time of the fusion, as all the textbooks describe; but the egg is *not* a haploid 'ovum' at the time that the sperm first makes contact. It is still an oocyte – still diploid – but is arrested in MII. The egg completes meiosis, expelling chromosomes surplus to requirements in the form of the second polar body, only *after* the sperm has made contact. Note, once more, that the initial contact and subsequent penetration of the sperm achieves two conceptually different results. First, contact with the sperm promotes the oocyte to complete meiosis; and once meiosis is complete then the cytoplasm of the oocyte may continue to divide, virtually without reference to the nucleus it contains for, as we have already seen, it is the *cytoplasm* of the newly formed zygote, not the nucleus, that controls the first few cell divisions. The prompting of the oocyte to complete meiosis and the initiation of embryo formation are collectively called 'activation'. In practice, embryologists sometimes apply the term 'activation' just to the completion of meiosis; and sometimes apply it to the completion of meiosis plus the first few cell divisions in the new embryo, in which embryogenesis gets under way. Quite separately, and in addition to activation, the penetration of the sperm provides new genetic material; and this is what truly counts as 'fertilisation'.

The sperm is, in effect, a travelling nucleus: it arrives in the oocyte cytoplasm complete with nuclear membrane (or nuclear envelope – 'NE') but sheds its tail after penetration and sits within the oocyte as a little dark sphere. It then swells – for the nuclear material is highly compacted within the sperm – and so becomes a pronucleus. Since the oocyte chromosomes are in mid-meiosis when the sperm arrives they have no nuclear envelope; but once they have completed meiosis they acquire a new nuclear envelope and then they, too, form a pronucleus.

For a brief interval, the two pronuclei sit side by side in the cytoplasm. The male pronucleus is already in a state that can be called G1 while the female pronucleus, once its new nuclear envelope has formed and its chromosomes have decondensed after finally completing meiosis, also enters G1.

Then both pronuclei replicate their DNA; or, in other words, they both enter S phase. But their entry into S, and S itself, are not tightly coordinated one with another. The two pronuclei remain roughly in step, but each takes its own time. For the moment, though side by side within the zygote, they are still quasi-independent beings. After S, each pronucleus is diploid; so the zygote as a whole (as such it now is) is tetraploid, since it contains two diploid pronuclei. In its diploid state, each pronucleus is of course in G_2.

At the end of this first G_2 the two pronuclei, side by side, enter mitosis: their first M. As always in mitosis, their nuclear envelopes break down, their DNA condenses into chromosomes, and so on. A single spindle forms within the cytoplasm. Only at this stage do the chromosomes form the two pronuclei, which before had not been tightly coordinated, begin to operate as a team. Both sets of newly condensed chromosomes respond to the call of the spindle and line up along its central plane; and from now on mitosis proceeds as in any other cell. So now there are two sets of homologous chromosomes lined up along the spindle, one derived from the male parent and one from the female, and each chromosome is a doubled structure: two chromatids joined at the centromere. Now they all split apart. The chromatids, once separated from their twins, can now be called chromosomes. So we finish up as always with two complete sets of chromosomes – though each one is now a single and not a doubled structure. Each complete new set of chromosomes now acquires its own nuclear envelope, to form a true nucleus; and each nucleus contains a complete set of chromosomes that originated in the sperm of the father, and a complete set that derived from the egg of the mother.

Since mitosis has taken place we now have a two-celled embryo. Intriguingly, it is only at this stage – in the two-celled embryo – that sets of maternal and paternal chromosomes come together within a single nucleus, and begin at last to operate as a coordinated chromosomal team. The notion that the zygote is itself a new individual is somewhat compromised by the biological details. The zygote contains a complete complement of chromosomes to be sure; but within the zygote, at least until mitosis gets under way, these chromosomes remain resolutely separate, with half in the maternal pronucleus, and half in the paternal pronucleus. They do not work together at the zygote stage; and, except perhaps in mice, the individual genes remain effectively non-functional, since they are not transcribed at this stage. The stage that we associate with independent life – when a complete complement of genes are not only working together, but are actually producing proteins – is not realised in humans, sheep and cattle until the eight-cell stage, and in frogs not until the embryo has around 4000 cells. This is an extraordinary biological fact; and it surely has significance for theologians and moral philosophers, seeking to define the point at which a new individual begins.

These, however, are just the background points: the basic terms. Now we come to the part that really matters: how the cell cycle is controlled. There is mitosis; and there is DNA replication (S phase). The DNA must replicate completely, once and only once. Mitosis must not begin before S is complete.

So how is the cell cycle controlled? The explanation that follows is complicated, but it is as simple as it can be without being deceptive. Life is complicated; and the cloning of Megan and Morag, and then of Dolly, depends on attention to life's details.

The control of the cell cycle

The experiments that have shown how the cell cycle is controlled – although understanding is still far from complete – have been undertaken since the 1960s and are marvellously ingenious. In one set of experiments – before I was at Sussex – scientists removed the membranes from the heads of sperm, so that, in effect, only raw DNA remained; then added this DNA to fluids extracted from the cells of *Xenopus*; and found that the membraneless sperm heads acquired new nuclear membranes. Clearly, some factor in the frog cell extract stimulated membrane formation. Thus, in a test-tube, the scientists had imitated the events at the end of mitosis when each newly divided set of chromosomes acquires a new membrane and so forms a new nucleus. They then showed that the sperm-head DNA, now enclosed within a new membrane, went into S phase. In other words, it replicated. Furthermore, it replicated once, and only once. Thus the scientists had imitated the events that occur in the middle of interphase.

In general, then, the cell cycle is controlled by factors emanating from the cytoplasm; and by the kinds of experiments just outlined, and many more, biologists have gained some insight into what those factors are, and how they operate. This of course was my own concern at Sussex, and subsequently. Roughly, and as simply as possible, the picture seems to be as follows.

First it seems that the cytoplasm produces factors which are generically called 'licensing factors', which find their way into the nucleus and cause the DNA to replicate. In other words, these licensing factors stimulate the chromosomes to enter the S phase.

But if licensing factors are present in the cytoplasm, why doesn't DNA replicate all the time? Why does it replicate just once, during the S phase, and only once?

The reason is that the membrane that surrounds the nucleus – commonly known as the 'nuclear envelope' – acts as gate-keeper. So long as this is intact, licensing factors cannot enter. They gain access to the DNA only when the nuclear envelope is broken down – which happens naturally, as we have seen, at the beginning of mitosis. It seems, then, that in the course of a normal cell cycle the nuclear envelope breaks down at the beginning of mitosis: licensing factors gain access to the chromosomes; and when the nuclear envelope re-forms at the end of mitosis the licensing factors are trapped inside. For some reason (the mechanism is not yet understood) the trapped licensing factors then stimulate just one round of DNA replication, and only one. Breakdown of the nuclear envelope – the crucial step that allows the licensing factors to enter – is generally abbreviated to 'NEBD' ('nuclear envelope breakdown').

According to this scenario, then, the nuclear envelope (membrane) provides the control. It prevents the entry of licensing factors except at the beginning of mitosis, and then it locks them inside the newly formed nucleus. So now we need to ask, what prompts the nuclear envelope to appear and disappear?

The answer to this – and to much else besides – lies with another agent in the cytoplasm known as *maturation-promoting factor*, or *mitosis-promoting factor*, or *meiosis-promoting factor*. It is fortunate that 'maturation', 'mitosis' and 'meiosis' all begin with 'M' because this means that whatever role this multipurpose factor happens to be playing it can simply be called '*MPF*'. MPF is a key player in the cloning story, and will feature throughout the account of Megan and Morag.

MPF first became known in the 1970s. Biologists found that if they took cytoplasm from a frog oocyte that was already mature – that is, was in MII – and added it to a young oocyte that was still attached to the germinal vesicle and was therefore in PI (prophase I), then the young oocyte would resume development. In fact the young oocyte 'matured' into an MII oocyte. So the biologists hypothesised that the MII oocyte must contain a factor that induced maturation of PI oocytes – hence the name 'maturation-promoting factor'.

Note that an MII oocyte is not in interphase: it has been arrested in the course of meiosis. In short, the MPF came from a cell that was in the course of dividing. Did cells that were in the course of mitosis also produce MPF? It transpired that when extracts from *mitotic* cells were added to frog MII oocytes, they also matured. Furthermore, extracts taken from mitotic cells from bread yeast, or indeed from human cells, would stimulate maturation of frog MII oocytes. In short, it seemed that maturation-promoting factor is universal: *all* cells produce it in the course of cell division, and MPF from any species will operate in any other. As we have seen, this was the kind of research I was involved in at Sussex.

Then, as time went on, it became clear that MPF has a more general function. It does not simply prompt PI oocytes to mature into MII oocytes. MPF taken from dividing cells will prompt other cells to enter mitosis (or meiosis, if they are germ cells). This is why 'maturation-promoting factor' is also 'mitosis-promoting factor' or 'meiosis-promoting factor'.

In fact, it soon became clear that MPF promotes *all* the visible changes that we associate with meiosis and mitosis. It causes NEBD. It also causes the chromatin to condense at the beginning of mitosis, which is when DNA macromolecules float into view as chromosomes. It also prompts changes in the cytoskeleton so that the cell changes shape during mitosis. In fact the cell as a whole becomes more rounded – spherical – during mitosis and this enormously facilitates the final splitting of the cytoplasm, which follows the division of the nucleus.

So the mechanisms of control begin to fall into place. A rise in MPF in the cytoplasm causes the nuclear envelope to break down (which is what happens at the start of mitosis or meiosis). When the membrane breaks down, licensing factors from the cytoplasm gain access to the chromosomes. At the end of mitosis, MPF falls, and the nuclear envelope re-forms. This traps the licensing factors inside, and they prompt the chromosomes to replicate – in other words, to enter S phase. Replication happens once, and only once. MPF rises during G2 of the cell cycle, as the cell prepares for mitosis; and it falls during G1, just after mitosis. Once again, when contemplating nature, we should be stunned by its brilliance.

But what prompts MPF to rise and fall? This has become apparent only in the 1990s. It turns out that MPF is an enzyme of the kind known as a *'protein kinase'*. (The formal names of all enzymes end in '-ase'. A kinase is an enzyme that transfers a phosphate group from ATP (adenosine triphosphate) – which acts in living systems as the universal source of phosphate – on to a target molecule. But the details need not delay us.) This kinase in practice has two components. The major component is a protein known colloquially as *'cdc2'*. But cdc2 on its own does not do anything. It becomes functional – that is, it operates as a kinase enzyme – only when it is linked up with a second protein, of the kind known as a *'cyclin'*. It transpires that, whereas cdc2 is present in more or less constant amounts throughout the cell cycle, the concentration of cyclin that activates the cdc2 varies throughout the cycle. Thus the cyclin provides the fine control. It is an exquisite mechanism: we might say that the cdc2 operates as the engine of change, which is constantly on standby; and the cyclin is the switch that turns the engine on when required. When cyclin is present it combines with cdc2 to form the protein kinase that is otherwise known as MPF.

This, then, is how the cell cycle in general seems to be controlled. We can now link this description of the cell cycle together with the general discussion of reproductive controls as outlined in chapter 5. There we said that hormones (the particular hormone involved varies from species to species) stimulate the immature PI oocytes in the ovary to mature to MII, which is the phase at which they are ready for release. So how do these hormones operate? Well, directly or indirectly, they stimulate a rise in MPF – or at least, to put the matter in more detail, they stimulate a rise in cyclin, which joins with cdc2 to make MPF.

This story has one last episode. We noted in chapter 5 that oocytes are released from the ovary when they are in MII, and they stay in MII until the sperm makes contact (which normally occurs within the oviduct). So how do they manage to stay in MII? Because the concentration of MPF within them remains high. So long as MPF is high the chromosomes remain condensed, and the nuclear envelope is unable to re-form, and so meiosis is unable to proceed further. When the sperm makes contact, MPF falls, and meiosis runs to completion. The second of the two polar bodies (PBI) is shuffled off (and lodges in the vitelline space, beneath the zona pellucida).

So why and how does MPF remain high within the MII oocyte? What stops it from falling? Well, here we need to invoke one final controlling factor, known as *CSF*. CSF stabilises MPF; or, more precisely, it stabilises the cyclin, which joins with cdc2 to make MPF. So long as CSF remains high, then cyclin remains present, so MPF remains high, and the oocyte remains in MII.

And here is the final twist: CSF, in its turn, is sensitive to calcium. If calcium is present only in low concentrations, then CSF remains high; but if calcium rises, then CSF falls. So all you need to do to bring the oocyte out of MII, and allow meiosis to run to completion, is add calcium. The added calcium reduces CSF, and this reduces cyclin, which deactivates MPF, which releases the oocyte from its suspension in MII. Chain reactions of this kind are common in nature. The controls are exquisite.

This final part of the story – the role of calcium – has one implication which is of vital practical significance in Roslin's method of cloning by nuclear transfer. We noted earlier that the penetration of sperm during fertilisation has two quite different functions. First it 'activities' the oocyte, which means it prompts meiosis to run to completion, with the expulsion of the second polar body. How does it achieve this? By allowing calcium to enter – which sets off the chain of reactions outlined above, culminating in a fall in MPF.

Secondly, of course, fertilisation adds genetic material. But it is possible by artificial means to separate the two aspects of fertilisation: the activation and the addition of genetic material. The oocyte can be activated *without* adding new genetic material. Just add calcium to the oocyte and meiosis will run to completion. Contrariwise it is possible to add genetic material – a sperm or a nucleus from some other cell – without activating the oocyte. Just keep the oocyte in a medium that is lacking in calcium. Then when the sperm is introduced, or a nucleus is fused with it, there is no inrush of calcium and so the oocyte cannot mature, even though it has taken another nucleus on board. MPF remains high, and the receiving oocyte remains in MII, until calcium is added to the medium. Again, as we will see, this kind of manoeuvre plays an essential part in the Roslin cloning procedure. Donor nuclei can be added to cytoplasts that are high in MPF, and remain high; or high but falling fast (which is what would happen in nature); or low. The outcome of nuclear transfer can vary a great deal, depending on which of these courses is followed.

Of course, when I first began at Roslin (then IAPGR) in May 1991 this picture was far from clear. It began to clarify fully only as we completed the experiments of the early 1990s which led up to Megan and Morag, and then in the Megan and Morag experiments themselves. For me personally, the transition was far from smooth. The team I was to work with, under Ian, had started work in the previous October with two aims: to improve the efficiency of cloning from cattle blastomeres (embryo cells) and to isolate ES cells from cattle, as a route to genetic modification. The members of the team had already established their positions in the group. I came in late, with new ideas, but with some control over the science. I felt my position was resented. In fact I was unhappy, and even looked for another job: but I stayed because I wanted to work on cloning (and there aren't many jobs in cloning) and because I did not want to move house. My first daughter was born in June 1990 and, among other things, we just could not afford another move. I now know that Ian had also wanted to leave way back in 1982. Both of us, it seems, began the project that led to Dolly with baptisms of fire.

But I stayed, things got better, and Ian, Bill Ritchie and I began a series of experiments in nuclear transfer that were precisely along the lines I wanted to do. The idea was to add donor nuclei (karyoplasts) *at different stages of their cell cycle* to MII oocytes; and also – a second refinement – *to change the MPF status of the receiving oocytes,* to see how that affected the outcome. The results of those experiments in the early 1990s provided the theory – or at least reinforced the theory – and the technique that led to Megan and Morag. Ian, Bill and I summarised this phase of the work in 1993 in a paper in *Biology of Reproduction.*

Our immediate aims were twofold. First we wanted to see what happened when we transferred embryo nuclei, at various stages in their cell cycle, into recipient oocytes that contained different levels of MPF. I had an idea of what would happen, of course: no worthwhile experiment begins without a hypothesis. But we needed to see whether my ideas really stood up. Secondly – the practical aim – we wanted to improve the success rate of embryo transfer. We wanted to find out why transferred nuclei sometimes produced healthy zygotes that went on to become blastocysts, that in turn grew into healthy calves or lambs; why some reconstructed embryos developed only as far as the blastocyst stage, and then failed; and why some reconstructed zygotes failed altogether. From this knowledge we hoped to develop a protocol that was much more surefire.

Those early 1990s experiments were carried out in cattle, and we could not afford surrogate mothers: we were able only to produce blastocysts. If you are not trying to produce live births, then cattle do have advantages over sheep. First, we could obtain cattle ovaries from the abattoir, which would provide most of the material we needed: and we could do all the work *in vitro*. We could not acquire sheep ovaries from the abattoir even if we had wanted to (for reasons that have to do with the marketing of sheep meat), so if we had worked with sheep we would have had to get the cells we needed from live animals by superovulation (as described in more detail in the next chapter). It is obviously much easier for a whole raft of reasons to use tissues from animals that are already dead, wherever possible.

To do the experiments we needed young cattle embryos, to provide the karyoplasts; and we needed MII oocytes, to act as recipients (cytoplasts). We began with cow ovaries from the abattoir, and aspirated out the young oocytes, with their surrounding follicle cells. The oocytes at this stage are in PI; but one of the great advances of the 1980s (made at Cambridge) was to develop ways of maturing PI oocytes *in vitro*, to produce MII oocytes. Then, to produce embryos to act as nuclear donors, we incubated some of these MII oocytes with bull sperm – in other words, we carried out IVF. So we produced embryos; and then we employed yet another of those serendipitous agents that biologists employ to manipulate cells and tissues – one called nocodazole, which arrests cell development. So we stopped the embryos dividing at the four-cell or eight-cell stage. Since we arrested development just after the cells had completed mitosis, we knew they must be in G1. We also produced some embryos when their cells were in G2 by use of yet another agent, cycloheximide. None of this work would be possible but for the years of toil in other laboratories worldwide that provided the necessary techniques and reagents.

Other matured oocytes became the cytoplasts. Bill removed their nuclei (by the method first described by Solter and McGrath). Then he fused them *either* with blastomeres that had been arrested in G1, *or* with blastomeres arrested in G2. You can see immediately how these experiments in nuclear transfer differed from those reported in the 1980s: we knew exactly what stage the donor nuclei were at when we transferred them. This was the essential extra detail we were attending to.

We also studied the MPF levels of the receiving oocytes. You will recall that when an MII oocyte has just been released from the ovary it is high in MPF. It's the high MPF that keeps the oocyte in a state of uncompleted meiosis. But we showed that the MPF concentration falls very rapidly after activation. Within an hour of activation it has fallen to thirty per cent of its maximum level; and within two hours it has dropped to twenty per cent. In nature, then, we can assume that MPF is very high at the time of fertilisation, but that it drops very rapidly afterwards.

For our present experiment, however, we wanted to introduce donor nuclei (either in G1 or in G2) into cytoplasts that were either high in MPF, or low. To achieve the first of these aims, we simply fused karyoplasts and cytoplasts without activating the cytoplasts first – so the donor nuclei were plunged straight into an MPF-rich environment. To achieve the second, we activated the cytoplasts ten hours before we introduced the karyoplasts. By that time MPF had dwindled virtually to nothing.

So what happened? We'll describe the results first, and then discuss what we think was going on. When donor nuclei were put into a low-MPF environment, they simply sat there happily, and then after a time started to divide often going on to produce healthy-looking blastocysts. This was the case whether the nuclei were in G1 or in G2 at the time of transfer.

But when the nuclei were thrust into a high-MPF environment, the results were very different, and dramatic. First the nuclear membrane disappeared and the chromosomes within started to condense – as if they were about to go straight into mitosis. They did not go into mitosis, however. Soon (presumably as MPF levels started to fall), the nuclear envelopes formed again, and the chromosomes relaxed once more, losing their condensed appearance. In fact we expected this to happen: other biologists elsewhere had reported 'nuclear envelope breakdown' (NEBD) and 'premature chromosome condensation' (PCC) after nuclear transfer.

What really mattered, however, was the effect that these events had upon the transferred chromosomes. Those that were transferred while in the G1 state were unharmed. When their nuclear membranes re-formed, they entered S phase (doubling their DNA), and then the cells divided normally. But the nuclei that were in G2 at the time of transfer were severely damaged. They too went through another round of S after their nuclear membranes re-formed – and as a result they were then much too large. Embryos that had been constructed initially with G2 donor nuclei sometimes underwent a few divisions: after all, division in the early stages of development is controlled by the cytoplasm. But as soon as the genes were required to take over the controls – the time of genome activation – the embryos failed.

Aided by hindsight, we can see exactly what was happening. Nuclei placed in a low-MPF environment simply go on dividing as they would normally. Their nuclear membranes remain intact; they just continue as before.

But when the nuclei are transferred into a high-MPF environment the MPF strips the nuclear envelope away, and exposes the chromosomes within to all the rigours of the cytoplasm – including an inrush of licensing factors. When the nuclear envelopes re-form, the licensing factors are trapped inside, exactly as would be the case during mitosis.

So now we see why the fates of the G1 and G2 nuclei are so different. G1 nuclei are diploid. When the nuclear membrane re-forms around them and the licensing factors trapped inside, so what? The licensing factors prompt them to enter S phase – but they were going to go into S in any case; and there is a mechanism to ensure that when they are exposed to licensing factors they will replicate once and once only. So the extra licensing factors have no effect. They simply prompt the G1 chromosomes to do what they were going to do anyway.

Not so the G2 nuclei. The G2 nuclei have already been through S. They are already tetraploid. When they are exposed to a fresh burst of licensing factors they double again, and become octoploid. Some plant cells can tolerate octoploidy but for animals this is fatal. The overburdened nuclei might undergo a few subsequent divisions – under the control of the cytoplasm – but as soon as they are called upon to function, they fail.

Finally, we can assume that if nuclei were transferred while in the S phase, then they would suffer a similar fate to those in G2. Chromosomes during S are somewhere between diploidy and tetraploidy; so after untimely exposure to licensing factors they would be somewhere between tetraploidy and octoploidy. A frightful mess.

So what can we conclude? If we want to make healthy embryos

that will survive at least to the blastocyst stage, then we can do this by transferring nuclei, *at any stage*, into a low-MPF environment: that is, into MII oocytes that have been activated some hours before. In fact, low-MPF oocytes can be called *'universal recipients'* – a term we coined in the 1993 paper.

But we cannot transfer nuclei into a high-MPF environment with any hope of success unless they are diploid at the time of transfer. This means they have to be in G1. Either that – and this turns out to be the million-dollar point – or they can be in the 'quiescent' state, G0.

So, you might ask, if it is always safe to transfer nuclei into a low-MPF environment, why not do this? Why cause donor nuclei to run the gauntlet of the high-MPF environment? Well, if you simply want to produce blastocysts, then a low-MPF environment is good enough. But if you want to produce blastocysts that can be transferred into surrogate mothers, and go on to become live animals, then initial exposure to the rigours of MPF seems to be helpful.

But we will come to this. Suffice to say that our experiments of the early 1990s showed us what we needed to know: that the cell cycle of the donor cells matters; and the MPF concentration of the receiving cytoplasm matters; that low-MPF cytoplasts are 'universal recipients' which provide a welcome for donor nuclei at any stage of their cycle; and that donor nuclei cannot safely be placed into a high-MPF cytoplast unless they are diploid at the time of transfer – meaning G1, or G0.

These are the insights we took into our experiments in the winters of 1993–4, and 1994–5, which led to the births of Megan and Morag. For me, Megan and Morag are the stars: the pair that vindicate all the theory, everything I had been thinking about for a decade or more. Dolly is just the gilt on the gingerbread.

CHAPTER 9

Megan and Morag

Ian Wilmut and Keith Campbell:

Real science needs two ingredients: a hypothesis worth testing, and the means to test it. Research on the reproduction of sheep has to be done in winter, which is when they mate and conceive; and in the winters of 1993–4 and 1994–5 we had not one, but two very significant ideas to put to the test.

The first idea was the one that Ian had been working on since the mid 1980s: that to make viable embryos by nuclear transfer, you had to begin with nuclei that were still totipotent. In the early 1990s Jim McWhir was taking this idea to its logical conclusion, and seeking to produce, from sheep embryos, cultured cells that retained totipotency – comparable to, if not quite the same as, the ES cells of mice. In the end he did produce cells that clearly did retain some totipotency, which he labelled 'TNT4' cells – meaning 'totipotent for nuclear transfer'. More generally, we knew from previous experiments that cells change in culture – visibly, as well as in more subtle ways; and we wanted to see whether and how this change in morphology was reflected in loss of totipotency.

The second approach was the one advocated by Keith: the distinctly heterodox notion that differentiation in cells is *not* necessarily irreversible – that a genome in a differentiated cell can be reprogrammed and recover totipotency. It was (Keith said) primarily a question of ensuring that the donor nucleus, and the receiving cytoplasm, were both in the appropriate stage of their cell cycles, and were compatible with each other. The climax of the two seasons' work was the birth of Megan and Morag in the summer of 1995 – and was a triumph for the second, 'cell cycle' hypothesis.

In practice, we worked on Jim's hypothesis in the winter of 1993–4, and again at the end of 1994; and on Keith's idea in the early months of 1995. We reported the results of both pieces of research – let's call them the TNT4 experiments and the cell cycle experiments – in *Nature* on 7 March 1996 in a paper that was formally entitled 'Sheep cloned by nuclear transfer from a cultured cell line', although we commonly refer to it as 'the Megan and Morag paper'. Perhaps it was a mistake to describe both sets of experiments in one single account. The two underlying ideas are different and the report is short (*Nature* allowed only 1000 words for 'scientific letters' in those days) so it is not easy for outsiders to see what is going on. You can regard the account that follows as the long version.

Both seasons' work, and both methods, produced live, cloned lambs. Six were born in the first year (in the summer of 1994), and five in the second (in the summer of 95). The embryos that became Megan and Morag were produced in February 1995, and they were born in August 1995. These two truly began the new age of cloning – which means they initiated a new age of biotechnology. Dolly is more famous, but Megan and Morag were the real pioneers.

So far throughout this book we have mentioned details of technique in passing – McGrath's and Solter's method for enucleating cytoplasts; Willadsen's technique for protecting reconstructed embryos in agar, and so on – and at this point we should perhaps show how we borrowed techniques from these and other scientists, added to them, and fitted them all together. Science really is a collective endeavour, with everyone building on what was done before. But then, of course, *all* human culture is a collective endeavour.

First prepare your sheep . . .

At Roslin the cloning of sheep by nuclear transfer has become routine – but it remains one of huge and irreducible complexity. For a start it requires *four* different groups of ewes – all of which need to be prepared and treated in different ways – plus a few rams, who are involved at various stages. One group of ewes simply provides oocytes, which, when enucleated, become the receiving cytoplasts. Another group provides embryos whose cells are cultured to provide the karyoplasts, the donor nuclei. A third group act as temporary recipients, incubating the newly reconstructed embryos within their oviducts until they reach the blastocyst stage. The fourth group act as surrogate mothers – the blastocysts are transferred into their wombs where at least some of them develop and become lambs.

In such procedures the ewes that provide the karyoplasts should be of a different breed from those that supply the cytoplasts – so that when the lambs are born we and everyone else can see at a glance that they indeed developed from the transferred chromosomes, and not from chromosomes that were left behind at the time of enucleation. Of course the provenance of the lambs can later be checked by 'genetic fingerprinting', but it is good if the parentage is clearly visible as well. For example, in the Megan and Morag experiment, Welsh Mountain ewes provided the karyoplasts, while in these and in all the experiments the ubiquitous, local Scottish Blackface provided the cytoplasts (and also served as temporary recipients and surrogate mothers). Welsh Mountain sheep are middle sized and white while Scottish Blackface are slightly larger and of course have black faces.

Actors are warned not to work with children and animals and reproductive biologists who want an easy life might be advised in similar vein to stay well clear of sheep. Natural selection has ensured that they give birth only when their lambs are most likely to survive and in temperate latitudes, where winter and summer are so distinct, that means they give birth in spring – or, sometimes in the northern latitudes around Edinburgh, in early summer. Megan and Morag and Dolly were born somewhat later in the year than would be usual on commercial farms. Gestation takes about five months – it varies somewhat from ewe to ewe, but is typically around 147 days for Welsh Mountain and Scottish Blackface – which means that ewes generally conceive around Christmas or New Year: midwinter.

Although we both like working with animals the task of preparing the sheep falls to Roslin's excellent team of shepherds and technicians. If they want oocytes to act as receiving cytoplasts, or embryos to provide the donor karyoplasts, then they must first prime the ewes with hormone pessaries and injections to induce ovulation; and if they are preparing ewes to act as surrogate mothers then they must first treat them with other hormones to induce the state of 'pseudopregnancy' that will enable them to accept implanted embryos. If the ewes are to produce the embryos that will provide the karyoplasts, then the shepherds also run rams with the ewes. Then at the appropriate times the shepherds and technicians round up the ewes and prepare them for surgery – which is not an enviable task. Midwinter in the hills around Edinburgh is generally cold and often wet and although sheep are not exactly dangerous the hill breeds are agile, strong and alert, with no desire whatever to cooperate. For good measure, there are only about six grudging hours of light in a Scottish midwinter day so that much of the work is done in gloom. Sheep smell, too – of lanolin, the fatty exudate that waterproofs their wool. People know when you have been working with sheep. So the preparation and round-up can be cold, wet, muddy, hard, and smelly – but this is science just the same: no less than in those bright, warm, aseptic labs where scientists work with sweet and shining mice.

The operating theatre where the (minor) surgery is done has a unique atmosphere: farmyard meets hospital meets laboratory. The sheep come straight in from the damp outdoors but infection must be kept at bay. We work both with lowland breeds, like Poll-Dorsets and Finns, and with hill breeds, like Welsh Mountain and of course the local Scottish Blackface. Lowland breeds tend to be passive and agreeable – everyone's idea of a sheep – while hill breeds have to be tough and suspicious in order to survive. Their bellies must first be shaved and disinfected ready for the various operations, and the skin beneath the thick soft coat of wool is generally clean and pink. John Bracken manoeuvres the ewes in the 'firm but gentle' manner that all animal handlers must cultivate but they may struggle in any case, though they quickly calm down when injected with anaesthetic. Then they can be lifted on to the table like any patient in a clinic. Cattle have to be held in 'crushes', tilted on their sides, and ratcheted into place on what is essentially a fork-lift truck. Sheep may be harder to deal with than mice, physically speaking, but they are a lot easier than cows.

The surgery is carried out by various people: sometimes one of us, but mainly by Marjorie Ritchie, wife of Bill Ritchie. John Bracken handles the anaesthetics and monitors the pregnancies, and other technicians, shepherds and postgrads lend various kinds of hands. The mood is relaxed – very advisable if you have a dozen sheep to get through in a day. In operating theatres there is a wonderful dislocation between the serious thoughts that run through your head – How is the sheep responding to the anaesthetic? Any signs of distress? – and the chatter that keeps everyone going. Marjorie has a fine line in banter.

Colin Tudge:

Several different kinds of surgical operation must be carried out in the course of an experiment, so on any one day Marjorie (or sometimes Keith) might be gathering oocytes or embryos from donor sheep; or fitting the newly made embryos into the oviducts of temporary recipients; or removing blastocysts from recipients that were implanted with embryos the previous week; or transferring those blastocysts into the uterine horns of surrogate mothers.

To be more specific: ewes that are due to provide the oocytes which will become the cytoplasts, are first injected with gonadotrophin-releasing hormone (GnRH) to induce superovulation. On the following day – 28 to 33 hours after the hormone injection –

Marjorie (usually) flushes the oocytes from the ewes' oviducts. Flushing is done by surgery under anaesthetic. She inserts a syringe into one of the 'uterine horns' – the projections at the top of the uterus which lead into the oviducts – and flushes saline up through the oviduct. The oocytes are thus washed out of the top of the oviducts, where they are collected in a catheter. Hence the oocytes are flushed out of the oviducts from below – they are not sucked out from above.

Once the oocytes are gathered they are kept at the body temperature of a sheep – around 37 degrees centigrade – but in any case are enucleated within hours of recovery by Bill Ritchie or Keith. Bill is a very neat man, of military aspect. He wears clothes that fit rather than flop and he is a keen walker, especially over the 'Munroes', the 284 Scottish hills that are over 3000 feet (about 900 metres). 'Only another 184 to go,' he says. At the heart of his domain at Roslin is a wonderful binocular microscope with robot-like attachments that can hold a Swiss army knife assemblage of microdissecting instruments including glass needles, pipettes, and what you will. Each is controlled by screws and can be shifted with absolute precision, micron by micron, through three dimensions – up, sideways and away. The instrumentation is beautiful but it takes immense skill to handle it well: coordination, and a 'feel' for the ways in which delicate though sometimes recalcitrant biological material behaves when it is twiddled and prodded. Each hand at any one time has its own task; sometimes the same task changes from hand to hand; the cells must be shifted here and there, but kept in the plane of focus of the microscope. 'It's like driving a helicopter,' says Bill. Such concentration is very tiring: 'Three hours and you're knackered.' Biology is a cerebral pursuit but to an enormous extent the research may rely upon manual dexterity.

Although biologists speak of 'enucleating' the oocytes, at the time they are 'enucleated' they are in metaphase II – MII – and at this stage, of course, there is no proper nucleus. The nuclear membrane has broken down and the condensed chromosomes lie bunched together in the cytoplasm. The task, then, is not to draw out an entire nucleus but to scoop out a bunch of chromosomes. The most obvious complication is that the cytoplasm of sheep MII oocytes is opaque, and the chromosomes cannot actually be seen through the light microscope.

In practice, the oocyte to be 'enucleated' is first treated with cytochalasin B to soften the cytoskeleton, and with fluorescent dye that attaches to the DNA. Then it is gripped by suction to the end of a pipette, under the microscope. Nature, fortunately, has provided a convenient means for locating the chromosomes. Since the oocyte is in MII, this means that the first meiosis is complete. The first meiosis has produced a polar body, and this lies trapped beneath the zona pellucida. The oocyte chromosomes lie within the cytoplasm roughly beneath this first polar body. So, says Bill, 'I just have to bring this polar body into view, thrust the end of a pipette through the zona, and then apply suction to draw out the polar body plus the underlying cytoplasm – and this *ought* to contain the chromosomes. Then to make sure that I really have removed the chromosomes, I shift the embryo out of the microscope field of view and shine ultraviolet light upon the pipette. Like so. If the chromosomes are in the pipette then they will glow – because they have been treated with fluorescent dye. If there is no glow, then the enucleation has not been successful.' Thus it is that feats that seem impossible – removing the threads of chromosomes from a tough little cell the size of a speck – can be carried out routinely and reliably, albeit with techniques and instruments that have taken centuries to evolve. Bill places each oocyte that is successfully enucleated to one side, with its surrounding zona, and moves on to the next one. He can get through scores in a day.

To reconstruct embryos, Bill or Keith begin with a pile of enucleated oocytes – which now can be called cytoplasts – and a pile of karyoplasts that have been separated from the rest of the cell culture with the aid of enzymes. The cytoplast is now held at the tip of a pipette as before, and twiddled to reveal the hole in the zona that was made previously, when the polar body and underlying chromosomes were sucked out. Now another pipette is used to insert a karyoplast through the wound in the zona, so that it is in contact with the outer membrane of the oocyte cytoplasm. All that remains is to place the two cells, now bound loosely together by the zona, between a couple of electrodes and apply the electric current – the electrofusion method which Steen Willadsen first used for embryo reconstruction in the mid 1980s.

After fusion each 'reconstructed embryo' is coated in agar for protection (another Willadsen innovation, which he in turn derived from earlier work by other scientists on rabbits) and is ready to be implanted into the oviduct of a temporary recipient. Thirty to forty reconstructed embryos are put into just one of each ewe's oviducts; and the receiving oviduct is then tied off below the young, agar-coated embryos to create a kind of incubating chamber, where they should be easy to find after they have developed a little. All this – embryo reconstruction, embedding in agar, and transfer to the temporary recipient – takes place within 50 to 54 hours of the initial GnRH injection.

Ian Wilmut and Keith Campbell:

Six days after we place the embryos into the temporary recipients we – usually Marjorie – retrieve as many as possible. By then, at least some of the reconstructed embryos should have developed into blastocysts. Then they go to Keith or Bill who prise them loose from their agar cocoons – which Bill says is one of the most difficult bits of the whole operation. Then we (generally Bill or Keith) examine them under the microscope to see which have developed into morulae or blastocysts. Those that have developed are transferred into the uterine horns of the 'permanent recipients' who thus become the surrogate mothers – assuming that any of the pregnancies actually 'take'. In general, we transfer two blastocysts into each surrogate mother. If two are present, the chance that one will be accepted is enhanced and it should be no problem if both of them implant and develop since ewes are perfectly able to produce twins.

The entire suite of experiments has to be fitted into the breeding season – which lasts at best from October to March. All the necessary procedures need to be dovetailed. Tricia Ferrier (a pleasantly harassed mother of one) calls herself the 'gardener of cells' and has to be on hand with the cultured karyoplasts; Bill Ritchie and

Keith Campbell need to be ready to reconstruct the embryos as the oocytes become available; temporary recipients have to be prepared to receive the reconstructed embryos and surrogates have to be in the right condition to receive the blastocysts that should develop from those embryos six days later. Typically, ewes are injected with GnRH on Monday morning, the oocytes are recovered from them on Tuesday morning, the embryos are constructed on Tuesday afternoon and evening, and they are then placed into the temporary recipients on Wednesday morning. The following Tuesday morning the embryos are recovered from the temporary recipients. Tuesday, then, is a busy day: first thing in the morning embryos are recovered from the previous week's implantation; and these are allowed to recover in culture before those that have developed normally are transferred to recipients in the afternoon. In between times, during the rest of the morning, the oocytes are recovered from another batch of ewes. It must be very difficult for anyone not directly involved to see what is going on – like trying to make sense of the Stock Exchange. The work also has to be done within budget, which is a tedious point but nevertheless crucial, and is one reason for working with sheep rather than cattle. In addition, because duration of pregnancy is shorter in sheep and they mature at a younger age, experiments can be completed more quickly. So we have a convenient animal to study for which there are also real benefits of cloning and genetic modification.

Once the embryos have been put together and duly implanted in their surrogate mothers, we just have to wait for five months with our fingers crossed while – with luck – the pregnancies run their course. John Bracken monitors the surrogate ewes with ultrasound – every two weeks at first, and then more frequently – to see if the ewes are indeed pregnant, and how the fetuses are faring. It is always a tense time. The whole year's work depends on these pregnancies. In fact the whole course of our research depends on them. Our entire careers. Indeed – dare we say – a large part of all biotechnology. Sheep have obstetric problems at the best of times (shepherds say that sheep spend their time thinking up new ways to die), and pregnancies that begin with cloned embryos are obviously more precarious than usual. Mice breed all year round and their pregnancies last only three weeks.

No wonder most biologists prefer them!

So far, however, we have described only half of the operation. The key to our experiments in the winters of 1993–4, and 1994–5, lay in the way we prepared the karyoplasts – the cells that provided the donor nuclei.

Cells for transfer

In all cloning experiments carried out before the Roslin work of the 1990s, cells would simply be separated from an embryo and fused directly with the cytoplasts. But the key point of our experiments of 1993–4 and 1994–5 was to grow and multiply the donor cells in culture *before* constructing embryos from them. Jim McWhir, working with Ray Ansell, devised an extremely efficient method for doing this.

They began with nine-day-old embryos – in other words, embryos who were flushed from the oviducts of donor ewes, nine days after mating. At this age a sheep embryo is an advanced blastocyst: the cells of the inner cell mass (ICM) have already formed themselves into an *'embryo disc'*, while the trophectoderm (TE) cells remain clearly separate, ready to make first contact with the wall of the uterus after implantation. Jim had already established that at this stage the embryo disc provides the maximum number of ICM cells that still retain a high degree of totipotency.

Jim and Ray then cultured cells from the embryo discs. They placed the embryo discs in flasks that had already been lined with 'feeder' cells – which in this instance are mouse fibroblasts – in a medium which among other things contains fetal calf serum, 'newborn' serum, and recombinant (genetically modified) human *leukaemia inhibition factor (LIF)*. The purpose of this arcane environment is to make the cells of the sheep embryo discs feel comfortable and multiply, while the LIF discourages differentiation. The fetal calf serum provided vital 'growth factors' (although exactly what these factors are and how they work is not precisely known).

For these particular experiments, Jim and Ray first cultured the embryo disc – 'ED' – cells for five to seven days, then lifted them from the substrate with trypsin enzyme, and then divided them to form four new cultures. They repeated this procedure up to thirteen times – so there were up to thirteen 'passages'. Cells may undergo all kinds of changes in culture, particularly as they are passaged; they may differentiate, mutate, or divide unequally and so become aneuploid. At first, both in the 1993–4 and in the 1994–5 experiments, the cultured ED cells looked rounded and undifferentiated, just like mouse ES cells; it seemed for a while as if Jim's and Ian's dream of producing ovine ES cells had been achieved. But, in all the cultures, by the second or third passages the cells flattened themselves against the substrate – so that they looked more like fibroblast cells of skin. Such cells look relatively unspecialised but they do show definite signs of differentiation. In particular, by the sixth passage they start to produce a couple of proteins (called 'cytokeratin' and 'nuclear lamin A/C') which we regard as 'markers' – signs that differentiation has begun. In the TNT4 experiments the cells were simply taken from culture after various numbers of passages and transferred into cytoplasts to make embryos. But in the cell cycle experiments, carried out early in 1995, the karyoplasts were put into the quiescent state – Go – before transfer; and how this was done we will discuss later.

These, then, are the basic procedures: how we make embryos; and how in general Jim and Ray cultured the cells that provided the karyoplasts. So now we should describe the actual experiments in more detail. We will begin with the experiments that effectively ignored the cell cycle of the karyoplasts, and contrived instead to culture them in a totipotent state.

The 'totipotent' hypothesis: 'TNT4'

This experiment began in the breeding season of October 1993 to February 1994, and continued into the next season (October to December 1994). As we have seen, the culture medium contained a factor specifically intended to inhibit differentiation – namely LIF: recombinant human leukaemia inhibition factor. But after two or three passages, the cultured ED cells clearly *had* differentiated, and indeed, as the *Nature* paper recalls, had 'assumed a more epithelial, flattened morphology'. They kept this shape at least until passage twenty-five. They differed from ES cells of mice both in appearance (ES cells are more rounded) and also biochemically – for at passage six they expressed two proteins that are associated with differentiation: cytokeratin and nuclear lamin A/C.

In the initial breeding season, we first constructed embryos using cells taken directly from sixteen-cell embryos as controls, and then made embryos from cultured ED cells that had been through one, two or three passages. At this stage the cells had already assumed an epithelial form, but were not yet expressing the two marker proteins. We reconstructed ninety-two embryos of which fourteen survived to the morula or blastocyst stage, when they were suitable for transfer into surrogate mothers. We transferred six of the embryos made with nuclei from sixteen-cell embryos into surrogate mothers, and two of these went to term. A single embryo made from an uncultured ED cell failed to make it. But one out of four embryos that we made from a first-passage cultured ED cell became a live lamb; the single embryo made from a second-passage ED cell also succeeded; and so, finally, did both of the two embryos made from third-passage ED cells, when they were put into surrogate mothers.

In these experiments, all the karyoplasts were transferred into cytoplasts that had been activated some hours previously, so that MPF was low (as detailed in chapter 8). We made no attempt at this stage to adjust the cell cycle of the donor nuclei – because, if MPF is low, this is not necessary; low-MPF cytoplasts are 'universal recipients'. Even so, six lambs were successfully cloned – two of them from ED cells that had been cultured to the third passage. Evidently, these cells retained totipotency. Jim McWhir described these cells as 'TNT4' – short for 'totipotent for nuclear transfer'. Accordingly, in 1995, TNT4 cells became the subject of a patent (– application number PCT/GB95/02095) which bore the names of J. McWhir and K. H. S. Campbell.

The following season – October to December 1994 – we continued this first set of experiments – but this time we used cultured ED cells that had gone through more passages: at least six, and up to eleven. We prepared 218 embryos, of which thirty-three became morulae or blastocysts which we transferred into surrogate mothers. But this time we had no success. Of nine embryos that we prepared from sixth-passage ED cells only one led to a pregnancy when implanted into a surrogate mother, and that aborted after 70 to 80 days. None out of ten embryos that we made from eleventh-passage cultured cells survived when put into surrogate mothers. Jim in particular felt that some extraneous factor might have been at work: after all, none of fourteen embryos prepared as controls from sixteen-cell embryos survived either – but these should have produced some live lambs. So we cannot infer the negative: we cannot conclude that cultured ED cells *cannot* retain the 'TNT4' state when taken beyond the third passage. But neither can we claim the positive. These experiments provide no evidence that ED cells will remain 'totipotent for nuclear transfer' when taken beyond the third passage. Perhaps the differentiation that is evident at passage six with the production of the two particular 'marker' proteins (cytokeratin and nuclear lamin A/C) does represent a critical barrier after all.

So this was the TNT4 part of the experiment. It shows that karyoplasts taken from cultured ED cells up to the third passage, whose nuclei have not been 'coordinated' by putting them into Go, can nevertheless produce live lambs when put into enucleated oocytes with low MPF; in other words it seems that, for at least the first three passages, some at least of the cultured ED cells do retain totipotency. Six cloned lambs born in the summer of 1994 suggest that this is the case, which is not a trivial finding.

It is overshadowed, however, by our second set of experiments, carried out from January to March 1995.

The cell cycle experiments

As described in the last chapter, our experiments in the early 1990s showed how the fate of transferred karyoplasts depends very much on the MPF content of the cytoplasts. If nuclei are diploid – meaning in G1 or Go – then they can safely be transferred whether the MPF content of the receiving cytoplast is high or low. But if they are not diploid – if they are in S or G2 – then they cannot be transferred safely into a high MPF environment. They can, however, be transferred into low-MPF environments without damage. This is why low-MPF cytoplasts are termed 'universal recipients': they can accept karyoplasts at any stage of the cell cycle, without damaging the chromosomes.

However, Keith also suspected from the beginning that a high-MPF environment might, in the end, produce better results. He reasoned that if the transferred nuclei are differentiated at the time of transfer – as they would be if they come from any tissue apart from a very young embryo – then their genomes need to be reprogrammed before they can support development into whole new embryos. The mechanism by which this reprogramming occurs is not understood in detail. It does seem, however, that reprogramming is likely to require exposure to factors in the cytoplasm. MPF in the receiving cytoplasm strips away the nuclear membrane and exposes the chromosomes to these (hypothetical) reprogramming factors. So, although it is safer in the short term to transfer nuclei into low-MPF cytoplasts, we also wanted to transfer nuclei into high MPF cytoplasts. This meant that the donor nuclei had to be diploid; and this meant that they had to be either in G1 or in G0.

These observations defined the structure of the cell cycle experiments, as carried out in January and February of 1995. First, we had to ensure that the donor nuclei were diploid at the time of transfer. Secondly, we produced a range of cytoplasts in three different states: some in which MPF was high, and remained high after the donor nucleus was introduced; some in which MPF was high initially, but fell rapidly after transfer; and some – the universal recipients' – in which MPF was low at the time of transfer.

To fulfil the first requirement – ensuring that the karyoplasts were diploid – we simply put all the cells into the quiescent state, G0, before transfer. This is remarkably easy, because cultured cells become quiescent if they are deprived of growth factor. Growth factor is contained in the fetal calf serum, so we just had to reduce the concentration of serum for seven days before transfer, from ten per cent (one part in ten) to 0.5 per cent (one part in 200). The newspaper reports that followed the births of Megan and Morag and of Dolly generally said that the cells were 'starved' to put them into Go, which implies withholding of nutrients; but, in fact, we withheld growth factors. Under such privation the cells that are in GI enter Go; and those that are in any other stage of the cycle either die or continue to cycle until they reach GI, and then go into Go. Thus at the end of the deprivation period all the cells should be in Go. At first Keith thought that putting the cells into Go 'was just a lazy way of ensuring they are diploid. It is much easier to put them into Go than into GI.' As we will see, however, he came to realise that Go is *not* just a variant of GI, but has special and necessary qualities of its own.

To prepare cytoplasts of the three different types – high MPF, high-but-falling MPF, and low MPF – Bill and Keith simply decoupled the two components of conception, which, in nature, tend to be conflated. For, as we saw in chapter 5, natural fertilisation has two outcomes. As the sperm penetrates the egg it obviously adds more genetic material, which of course is the whole point. But as the sperm makes contact, it also activates the egg – inducing the MII oocyte to complete the second meiosis, and shuffle off the second polar body. This occurs because the entry of the sperm allows calcium to rush in, which causes MPF to fall; and once MPF falls, meiosis goes to completion.

Addition of a karyoplast in effect mimics the addition of a sperm. Of course, an MII oocyte in a state of nature still contains its own chromosomes, whereas an MII oocyte used for embryo reconstruction has had its chromosomes removed; but the cytoplasm behaves the same way whether it contains chromosomes or not. We showed in the early 1990s that it is possible to activate an egg, and so to cause its MPF to fall, *before* adding the karyoplast. This can be done by exposing it to a suitable electric shock. In the 1995 we prepared low-MPF, 'universal recipient' cytoplasts by activating them four to six hours before cell fusion. Thus, by the time the karyoplasts were introduced, the MPF content of the cytoplasts had already fallen.

But we can also activate the cytoplast at the same time as the donor nucleus is introduced. Then it has a high MPF content at the time of fusion – but this falls rapidly. This procedure most accurately reflects normal conception. Keith invented an acronym for this second procedure: *GOAT*, meaning '*Go activation and transfer*'.

Finally, we were able to fuse karyoplasts with cytoplasts *without* activating the receiving cytoplasts. In such cases the MPF remains high, and the chromatin of the introduced karyoplasts are exposed to MPF until activation is achieved – which, in our 1995 experiments, was about six hours later. You may wonder how a nucleus can be transferred into an oocyte without activating it, since it seems as if activation would be achieved simply by penetrating the cell membrane, which is what the sperm does in a normal fertilisation. But, as we saw in chapter 8, it is not the penetration *per se* that brings about activation: it is the influx of calcium that accompanies penetration. Calcium deactivates the CSF, which is the agent that sustains the level of cyclin, which is the essential component of MPF. So you can effect cell fusion without activating the oocytes if you do this in a calcium-free medium. No calcium, no influx, no activation. Activation can then be achieved electrically as required, in the presence of calcium. Keith had another acronym for this procedure: *MAGIC*, meaning '*metaphase-arrested G1/G0 accepting cytoplast*'.

To summarise: universal recipients are activated some hours before fusion with the karyoplasts, and so are low in MPF; GOAT cytoplasts are activated at the time of fusion so their MPF content is high at first but falls rapidly thereafter – and this is the state that most closely imitates nature; and MAGIC oocytes are activated some time after fusion, and MPF remains high until they are activated. MAGIC and GOAT are handy terms for complicated procedures. We use them around the laboratory. But they do not unfortunately appear in the *Nature* report of 1996 because the paper's referee wanted them removed. Referees are not always right.

The cell cycle experiments differ from the TNT4 experiments in three outstanding ways. For one thing, the karyoplasts in this case had been through at least six, and up to thirteen, passages – so we might say they began where the previous experiments left off. Secondly – and, it seems, vitally – we put all the cells into G0 before we transferred them. Thirdly, each karyoplast was fused with a cytoplast that had been prepared in one of the three different ways: universal recipient, GOAT, or MAGIC.

In the first months of 1995 Keith and Bill Ritchie prepared 244 embryos from G0 karyoplasts of which thirty-four developed to morulae or blastocysts suitable for transfer to surrogate mothers. Ten of these survivors had been prepared by the MAGIC technique – placing the G0 karyoplasts in a high-MPF cytoplast – and we transferred these into six ewes. We produced sixteen embryos suitable for transfer into surrogates by the GOAT protocol, and distributed these among nine surrogate ewes. We made eight embryos suitable for transfer into surrogates by fusing G0 karyoplasts with 'universal recipient' cytoplasts that had been activated several hours earlier, and were low in MPF. We distributed these embryos between four surrogate mothers.

Then we held our breath. If these experiments worked – making embryos from the nuclei of cells that were clearly differentiated – then we knew we would have taken the whole science and craft of cloning, and indeed of biotechnology as a whole, on to a new plane. We had other things to do that spring (you always have to be thinking ahead to the next round of experiments; writing applications for grants; going to meetings – a hundred things) but our thoughts focused around the ewes that carried (or did they?) this quite new class of creature: animals cloned from cultured, differentiated cells.

John Bracken scanned the ewes with ultrasound and, by March, he was detecting pregnancies. One of the six ewes given MAGIC embryos was carrying twins. One out of six might not seem tremendous; but with this kind of novel experiment we were well satisfied. No fewer than five of the nine who were given GOAT embryos were pregnant; each carried a single lamb. One of the four ewes carrying an embryo made with a 'universal recipient' cytoplast was also carrying a lamb. This was very satisfying. It meant that all three procedures had produced pregnancies.

Colin Tudge:

That same month, March 1995 – with the ewes in the crucial experiment already pregnant – Keith had a revelation. Up until then he had generally assumed that donor nuclei should be put into G0 just as a way of ensuring that they are diploid – simply because diploid chromosomes are undamaged by transfer into a high-MPF environment and a high-MPF environment seemed necessary for reprogramming. But in March 1995 he finally realised that there was probably a great deal more to G0 than this. G0 is called the 'quiescent' state – but mere quiescence is not the point. G0, it seems, is a special state in which chromosomes are particularly amenable to reprogramming; indeed it seems that *all* cells pass through a G0 state *en route* to differentiation. ES cells of mice do not go into G0 – so long as they remain as ES cells. But they do enter G0 when they are induced to differentiate – which they do when they are placed in the inner cell mass of an embryo.

At least, Ian says that Keith realised the special significance of G0 in March 1995, when it was first confirmed that the ewes carrying the G0 embryos were pregnant. Keith says he is not quite sure when that particularly weighty penny finally dropped. 'I suppose I'd been thinking about it for a long time,' he says. 'But in March 95 I realised we were likely to get live lambs – and I also realised that if G0 really was special, then we could patent it. So I knew we had to keep quiet about it until the patent went in. So that was when I put my thoughts on paper. I wrote Ian a memo – effectively saying we were on to something big, and we should keep quiet about it.'

Ian Wilmut and Keith Campbell:

That memo still exists in the Institute's safe. Although when we were writing this book neither of us could find a hard copy, Keith was able to summon forth a text from the depths of his computer. It emerged with a headline in thirty-two-point sans serif:

CONFIDENTIAL – MEMORANDUM

It goes on:

TO: Ian Wilmut. FROM: Keith Campbell.

CC: [*blank – for nobody else at all was intended to see this message!*].

DATE: 16th March, 1995.

SUBJECT: Potential patent application.

TITLE: Quiescent cells as nuclear donors.

The memo first points out innocently enough that 'Quiescent cells are diploid'. But then it adds – 'The Go state is the point in the cell cycle from which cells are able to differentiate'.

Keith then lists a catalogue of physical changes which suggest that Go really does represent a special state. We need not explore the underlying ideas but you will surely get the gist: 'On quiescence a number of changes have been observed in quiescent cells. These include: monophosphorylated histones [these being the proteins that form the core of the chromosomes], ciliated centrioles [the entities round which the spindles form in mitosis], reduction or complete cessation in all protein synthesis, increased proteolysis [breakdown of proteins], decrease in transcription and increased turnover of RNA resulting in a reduction in total cell RNA, disaggregation of polyribosomes, accumulation of 80 S ribosomes, chromatin condensation.' These details, thus listed, do not by themselves show how the Go state contributes to the programming of genomes. They do suggest, however, that cells in quiescence are not simply sulking, but that Go represents a very particular set of manoeuvres; and other observations suggest that these changes do contribute vitally to differentiation.

So then Keith's memo asks, 'WHY IS THIS IMPORTANT FOR NUCLEAR TRANSFER? – and provides the answer: 'Many of these features are those which are required to occur following transfer of a nucleus to an enucleated oocyte. The fact that the Go state is associated with cell differentiation suggests that this may provide a nuclear/chromatin structure which is more amenable to remodelling/reprogramming.'

In short, Go is not just a lazy way of ensuring that karyoplasts are diploid. Go has its own, special status. It is a necessary step on the path to differentiation. In the context of cloning technology, it could be a crucial stage in genomic reprogramming – the means to reinvoke totipotency in differentiated cells.

So we did keep quiet about the methods that had produced Megan and Morag. From the time of that memo in March, to August when Megan and Morag were finally born – and indeed until the patents were safely lodged – we, and the other authors of the *Nature* paper, said nothing to anybody, not even Roslin's director. This made for a strange atmosphere within the lab. Other people knew that we had done something significant, but we wouldn't say what. So we would have lunch with friends and colleagues in the canteen, but we just wouldn't talk about the work that concerned us most. Scientists of thirty years ago would have thought this was dreadful. The whole point and pleasure of working in a lab was to share ideas. But these are hard commercial times. Sometimes you just have to hold your peace – however much you might want to shout about it!

Our anxiety grew as the weeks passed. We knew that if the ewes that were carrying the embryos made from late-passage, Go karyoplasts did finally give birth to live lambs, then one of the most deeply entrenched of all the dogmas in biology would be overturned: differentiated mammalian cells could be reprogrammed; mammals could indeed be cloned from such cells. As scientists, driven by the desire to understand life's processes, this was thrilling: to have contributed to a new phase of understanding. We also knew we would have at least one and probably two extremely valuable patents that showed how such mammalian clones could be made. On balance, Keith perhaps is more of a 'pure' scientist, above all wanting to know how life works; while Ian is happy with the idea that science should, somewhere along the line, produce something useful. But these are nuances.

Both of us were on tenterhooks for reasons both intellectual and commercial.

Keith Campbell:

Despite the folklore, too, sheep are not relaxing animals. They are notoriously prone to obstetric problems, which is why in the lambing season shepherds traditionally spend their nights out in the fields. So as the birthing hours came near, a group of us – John Bracken, Bill Ritchie, Douglas McGavin and I – took it in turns to sleep near the sheep sheds, looking in at the animals for any signs of parturition, every hour on the hour, from midnight until 5.00 a.m. The farm staff came on duty after that, and could take the six o'clock watch. For the first few nights I sprawled out in my caravanette next to the sheds but I hated having to get dressed and rush out through the cold night air every hour, so I soon transferred my sleeping bag to the office next to the sheep. It is never quite dark in these parts in midsummer but it could be pretty spooky all the same. I slept out in this up-and-down way for several nights each week for three weeks, and afterwards would drive home – it's about 16 miles – to snatch an hour or two of sleep before the children came in to say goodbye on their way to school.

I would like to report wild adventures along the way but absolutely nothing happened. The excitement came only from the rabbits who crowded on to the road on sunny mornings to warm themselves up, so that I had to slalom around them; and once when I went for a sandwich at the local all-night garage at around two in the morning the police followed me all the way there and all the way back, though only it seems because they were just as bored as I was. None of the sheep showed any signs of birth at all. In the end all the lambs were born in the afternoon, when I was on site in my office. The results justified all the effort, however.

The outcome: Megan and Morag

Ian Wilmut and Keith Campbell:

Here are the raw statistics:

> Of the six surrogate ewes who were carrying MAGIC embryos, one became pregnant with twins. One of the pair went on to become a live lamb but it died soon after birth. The live-born lamb came from a karyoplast that was in its eleventh passage.
>
> Of nine ewes who carried GOAT embryos, five became pregnant, each with single lambs. Three were born alive; one from a sixth-passage karyoplast; and two from eleventh-passage karyoplasts.
>
> Of the four ewes who were given embryos made with 'universal recipient' cytoplasts, just one became pregnant with a singleton lamb, which she took to term. The successful karyoplast had been in its thirteenth passage.

So we produced five live lambs in all – and each of the three protocols yielded at least one. But two of the lambs died within minutes of birth, and a third died at ten days. The other two, nearly five years later, are as healthy as ever. These are Megan and Morag. Marjorie Ritchie named them: the M is her own initial letter; Megan is Welsh, after the Welsh Mountain ewes who supplied their genes; and Morag celebrates the Scottish Blackfaces who provided the oocytes. Were it not for Dolly, these two young ewes would be the most famous sheep in the history of their species, and rightly so. When Morag was around 18 months old she became pregnant by a Welsh Mountain ram and produced a lamb, called Katy.

Megan and Morag in a sense are genetically identical twins, but their significance is greater than that. Analysis of their DNA – 'microsatellite' analysis – confirmed that Megan and Morag and all the other lambs who were born alive, plus the fetuses that did not make it to term, and of course the original cultured cells, all derived from a single population of cells. So all of them – the live lambs, the dead fetuses, the embryos that never made it, and the karyoplasts that never managed to turn into embryos – collectively formed, or form, one great clone. Megan and Morag are simply the two survivors from a much larger, genetically identical group.

So we have vindicated the 'cell cycle hypothesis'. When karyoplasts are transferred in Go they will produce live lambs even when they have been cultured for up to thirteen passages. However, this round of experiments did not support the notion that the Go karyoplasts are reprogrammed most effectively when placed in a high-MPF environment; indeed, Megan was born by the GOAT technique and Morag was made with a low-MPF, universal recipient cytoplast, while the one live-born lamb that had been created by MAGIC died soon after birth. But, as philosophers of science acknowledge, you should not abandon good ideas just because one set of experiments fail to fall exactly into line. There were too few animals in these experiments to show any statistically significant difference between the three protocols. Since Megan and Morag were born other scientists in other laboratories have produced clones from cattle and mice, and these later experiments do indicate that MAGIC on the whole gives the best results.

The Megan and Morag experiments also enabled us to apply for patents. One is titled 'Quiescent cells for nuclear transfer (1995), by K. H. S. Campbell and I. Wilmut. Application no. 9517780.4 (quiescent)'. The other has the title, 'Use of metaphase arrested oocytes as cytoplast recipients for nuclear transfer' (1995), K. H. S. Campbell and I. Wilmut. Application no. 9517779.6 (MAGIC)'.

A summary of cloning so far

The work we have described so far, at Roslin and elsewhere, suggests that scientists who affect to clone animals by culturing cells as nuclear donors, and transferring them into oocytes, should take at least six parameters into account. First, they should consider the *state of maturation* of the donor creature when culture begins – whether it is taken from a young embryo (before or after genomic activation), a fetus, an infant, or an adult. Then the *degree of differentiation* of the cell when culture begins is also relevant – whether it is highly differentiated, or is some kind of pluripotent stem cell. Thirdly, the *nature of the cell* when culture begins probably matters – whether it is a fibroblast, or comes from mammary gland, muscle, nerve, or what you will. Fourthly, the scientist must consider the *passage number* of the culture – how many times the cells have divided in culture, and how many times they have been replated. Then come two items that have to do with the cell cycle: the *state of the donor nucleus* – whether diploid or not diploid and, if diploid, whether in G0; and the *time of activation of the receiving oocyte* – before, during, or after cell fusion – and hence its MPF concentration.

Up until we made Megan and Morag, almost all biologists had thought mainly about the first four factors – all relating to the degree of differentiation of the donor cells. The whole endeavour to produce ES cells from farm livestock was based on the assumption that it was necessary to produce cells in culture that were differentiated only to the slightest degree. Our experiments in the early 1990s indicated the supreme importance of the fifth and sixth parameters, and particularly the fifth: and the birth of Megan and Morag show that this indeed is probably the most important concern.

But we would not want to suggest that the first four items on the list do not matter at all. John Gurdon's experiments way back in the 1960s and a great many observations since indicate that the degree of differentiation does make a difference, just as common sense suggests must be the case. Some body cells really might have differentiated beyond the point where totipotency can be recovered. All this remains to be found out, and the finding out will certainly be instructive. But, on present evidence, the supreme factor that seems to determine the success or failure of cloning more than any other is the stage of the donor cell cycle. G0, the state that seems to allow the genome to be reprogrammed, seems to be the key, while the state of the recipient cytoplast's MPF certainly seems to influence the outcome. The shift of emphasis, from degree of differentiation to cell cycle, is crucial and huge. This indeed is the insight that has made cloning feasible. As such it is one of the most significant insights of modern biotechnology. You may feel that such a claim is high-flown but we suggest that its importance can hardly be overestimated.

After Megan and Morag

As we published the formal paper in *Nature* Roslin announced Megan and Morag to the world at large. We and the institute as a whole thought that people should know what we had been up to – and, besides, we were proud of our success. Newspapers worldwide took it up. Keith learned of the publicity when he was still in bed one morning, and a friend called him up to say that we 'were in all the papers. So I went down to the newsagent's and found sheep on every front page.' Roger Highfield in the *Daily Telegraph* pursued the notion that, by cloning, women might reproduce without men: 'When sheep are sheep and men are uneasy', the headline ran. The headline over Robin McKie's account in the *Observer* was heavy on pun though low on obvious meaning: 'A clone again, naturally: Males don't figure in embryonic engineering, but it's not all ova yet'. Simon Houston and Chris Evans ran a good account in the *Daily Mail* on the very day of the *Nature* paper, albeit with a slightly lurid headline: 'Monsters or miracles'. But other events soon took over. In particular, a few days later, came the horrible shooting of schoolchildren at Dunblane. Understandably, this drove every other story from the newspapers – and once they are gone, they are generally gone. It is strange, though, the way that publicity for any one event – in this case a piece of science – depends on what else is happening in the world.

But life moves on. No sooner were Megan and Morag born – indeed before they were born – we had to be planning the next wave of experiments, to carry out in the winter of 1995–6. Megan and Morag had been made from cells that had been taken from nine-day-old embryos and had then been cultured, and had differentiated in culture. The fact that the donor cells had differentiated in culture seemed to override the fact that they had originally emanated from young embryo cells. If that was so, could we clone embryos from cultured cells that had come from older creatures – from fetuses, say, or even from adults? It had to be worth considering.

Thus it was that on 5 October 1995, just two months after the birth of Megan and Morag, we (in fact Ian) submitted an 'application for a research contract' to the MAFF. Under the box marked 'project title (maximum 120 characters)' Ian wrote: 'Development potential of quiescent cells derived from sheep embryos, sheep fetuses or adult sheep.' Under 'abstract of research' he made clear that 'developmental potential' was to be investigated by nuclear transfer. The total cost to MAFF, said Ian with commendable if somewhat implausible precision, would be £484 519.

This was our first formal reference to adult cloning. MAFF paid up and so began the research that led to Dolly.

CHAPTER 10

Dolly

Ian Wilmut:

Dolly has transformed my life. As the science and technology that produced her is swept up into the grand stream of biotechnology, she will touch everybody's lives. Of course as we waited for her to be born through the summer of 1996 – not even knowing whether any of the embryos we had made the previous winter *would* be born – I knew full well how important she was, and anticipated the impact she would make, although no one could have foreseen just what a fuss there would be or quite how frenetic life would become. I suppose my mood through that summer was half elated and half fearful – fearful that we would fail to produce a lamb like Dolly; but also, from time to time, fearful that we would succeed. We are private people, Keith and I, not heading instinctively for the world stage (although I have discovered a taste for it); and I knew that our lives would never be quiet again, and the world as a whole would be ever so slightly but ever so significantly different as a result of our endeavours.

What should we have done, then, when Dolly finally came into this world late on the afternoon of 5 July, after that decade of research and all those weeks of anticipation? People have not been short of opinions. Surely Keith and I should have painted Roslin village in a gaudy shade of crimson, or got uproariously drunk, or danced through the perpetual twilight of the Scottish summer night? Yet we didn't do those things. Perhaps my mind had switched off, anticipating the stress and freneticism that was to follow. The simple truth is that I can't remember what I did on the afternoon that Dolly was born. I did not attend her birth because John Bracken was in charge, and sheep are always liable to run into obstetric problems at the best of times, and the fewer people who hang around, the better. John says he thinks I was away digging my allotment when he telephoned to give me the news, but I honestly can't say whether I was or was not. I had bought a bottle of champagne to celebrate – assuming that celebration would be appropriate! – but Keith was away on holiday in Devon at the time of the birth (his partner is a social worker, and has to arrange holidays six months in advance) and although he was in constant touch by telephone it didn't seem right to open the bottle without him. I think I did nothing very much. I like to take our spaniel for a walk up the hill at the back of our house at the end of the working day. I probably did that.

In some ways, to be sure (as Keith is forever pointing out) Dolly is not quite so interesting scientifically as Megan and Morag. They after all were the very first mammals of all to be grown up from cultured, differentiated cells. Dolly may not have such practical significance as Taffy and Tweed, who were also born in the summer of 1996, but developed from fetal fibroblast cells, which for general purposes are probably the best bet. She has been superseded by Polly, who was born in 1997 and developed from fetal fibroblast cells that had already been genetically transformed. But Dolly has one startling attribute that is forever unassailable: she was the first animal of any kind ever to be created from cultured, differentiated cells taken from an *adult*. Thus she confutes once and for all the notion that has virtually been dogma for 100 years, which says that once cells are committed to the tasks of adulthood then they cannot again be totipotent. The cell that created Dolly came from an adult ewe – indeed the ewe that provided her genes was almost elderly – yet its ability to be reprogrammed into totipotency has been demonstrated beyond question. Keith was first to insist that differentiated cells could be reprogrammed, if proper attention is paid to the cell cycles of the karyoplast and the cytoplast; and this hypothesis is well and truly vindicated, startling and even ludicrous though many, just a few years earlier, would have thought it to be.

All in all, Dolly is the stuff of which myths are made. Her birth was other-wordly – literally a virgin birth; or at least, one that did not result directly from an act of sex. The scientific Magi were suitably amazed. Many had felt that cloning from adult cells was impossible and even the optimists had generally supposed that it would not be achieved for another 100 years. Even John Gurdon, the pioneer of modern cloning, was taken by surprise. As in all great myths, too, the nature of the birth was soon questioned: some scientists and journalists doubted whether Dolly was what we were claiming she was, or just some kind of charlatan – a mistake, or even a fraud. Some popular accounts accepted her reality but nevertheless suggested that she was simply an afterthought, not to say an accident, cobbled together at the last minute from some culture of cells that just happened to be knocking about the freezer, by implication wedged between a TV dinner and a packet of frozen peas. Thus, the myth had it (as myth always does), her birth was not guided entirely by human intent.

But Dolly is real, and she does derive her genes from an adult mammary gland cell – for all possible doubts on this were laid to rest by 'genetic fingerprinting' in the summer of 1998. Her birth certainly was not an accident and indeed we had long been thinking of cloning from adult cells although the way we finally launched into the work was in the end somewhat serendipitous (that dread but accurate word again!).

Just to recap briefly: Megan and Morag were born in August 1995; in the same month, Keith and I patented our novel methods for nuclear transfer; and, as described in the last chapter, Keith and I applied to MAFF for money to attempt cloning from adult cells in the October. In the end, however, the research in the winter of 1995–6, which led to Dolly, was a collaboration between Roslin and PPL; and the adult cells that produced Dolly, and some of the money to make her, came from PPL. For PPL possessed the vital ingredient that we lacked at Roslin: a line of cultured, adult sheep cells.

PPL had not acquired these cells in a fit of absent-mindedness as rumour seems to have it. They played a key role in PPL's pharming programme. PPL has survived and flourished because of Tracy and – even more importantly – the commercial flocks of transformed sheep that have followed in her wake. PPL were interested in cloning only as an adjunct to their work on genetic transformation. As we saw in chapter 3, it is not enough, in genetic engineering, simply to introduce novel DNA. The new gene has to be recognised by the host genome, and to respond to the signals of the host that control gene expression. Thus engineers cannot simply drop in DNA (even though they may talk casually at times as if they did). They first create gene 'constructs' in which the gene in question is joined to DNA promoters and other modifiers taken from the host. Tracy was made by attaching a human *AAT* (alpha-1-alphatrypsin) gene to the promoter that normally controls the production of a milk globulin in the lactating ewe. But there was no reason to assume that that particular promoter, or the 'construct' in which it was incorporated, were the best that could possibly be provided. In order to improve on Tracy – and to create other lines of sheep producing other proteins – the PPL scientists want to know exactly what genes are contained in the cells of sheep mammary glands, and exactly how they are all controlled. Then they can create constructs of genes that are expressed even more powerfully than the *AAT* gene in Tracy, and are under even tighter control.

For this purpose PPL has long been collaborating with the Hannah Research Institute, a government laboratory in Ayr about 70 miles to the west. The Hannah specialises in dairy science, and in March 1995 PPL and Hannah 'sacrificed' a six-year-old Finn-Dorset ewe who was in her last third ('trimester') of pregnancy and, under the direction of Dr Colin Wilde, they removed mammary tissue which they broke into fragments and froze. Mammary tissue contains several different kinds of cell, but the ones of main interest were '*ovine mammary epithelial cells*', or '*OME cells*'. Ever since 1995 Hannah and PPL scientists have studied these cells to see precisely what proteins they will produce, under what circumstances, what genes are responsible and what controls them. These are the cells that gave rise to Dolly. The idea that they were just spare, and vaguely knocking around half forgotten, is very far from the truth; they were, and are, key players in the research at PPL, very much in people's minds. They were not, however, initially intended as nuclear donors in cloning experiments.

The cloning of Megan and Morag in the winter of 1994-5 was entirely Roslin's initiative; but the research of 1995-6, which led to Dolly, was a collaboration between Roslin and PPL. We began making detailed plans for the winter's experiments in October 1995. Keith and I at Roslin wanted to clone from fetal fibroblasts, since these grow much faster in culture than the embryo cells that had produced Megan and Morag and seemed to be the most promising lines to exploit. PPL wanted to test two lines of embryo cells. But then chance took a hand. Cell culture is not an exact science and one of the lines of embryo cells cultured at PPL became increasingly aneuploid: as they divided, the cells lost or gained chromosomes. There is nothing to be done with such cells except throw them away. So this left an embarrassing gap in the winter programme. We had funds for four sets of experiments, we had bought the sheep and made our plans and suddenly one of the lines of donor cells, the *sine qua non*, was no longer available. At the start of October 1995 Keith was convinced that adult cells could be used and had in fact isolated a number of cell lines for this purpose. However, our government funding would not cover the cost of these experiments. In the collaboration with PPL excess funds were available due to the failure to isolate sufficient embryo derived cell lines. In addition, PPL were in possession of an adult derived cell line: the mammary gland cells that they had prepared with the Hannah, originally for a quite different purpose.

The final strategy was framed at a joint Programme Management Committee meeting at Roslin late in 1995. I chaired, and others present included Keith and Jim McWhir from Roslin, with Alan Colman, Angelica Schnieke (the 'g' in Angelica is hard, as in Anglo) and Alex Kind from PPL. Angelica Schnieke was a PhD student who was employed by PPL but had been working at Roslin, and although it was in Keith's and my mind it was she who first formally suggested that the mammary gland cells might be used to make clones. Indeed she gave several cogent reasons why this could be especially worthwhile. First, she argued, the

OME cells *looked* similar to the embryo cells that had given rise to Megan and Morag. Appearance – morphology – is a big thing in cell culture: if two cells' lines look similar, this suggests that they are responding to similar conditions in similar ways, which implies that the resemblances are not simply superficial. Secondly, the OME cells were precisely the kind that interested PPL most. Wouldn't it be good to create whole animals from the cells that were of most concern? That way – at least if the technique ever became established – the qualities of the newly cloned animals might be predicted from the properties of the very cells from which they were created.

So the idea of cloning from adult cells came initially from Roslin, and was long established: Keith had had it in mind for many months and we had applied to MAFF for funds in the autumn of 1995. Yet the experiment would not have been carried out that winter if PPL had not had a suitable line of cells which had been created for a quite different purpose: and they would not have been used if both of PPL's cultures of embryo cells had succeeded. Indeed, Alan Colman says that he would have vetoed the idea in his capacity as research director 'if PPL had been able to prepare more than one new embryonic line'. After all, he says, 'I had done my PhD with John Gurdon, and the failure of adult cell nuclei to make adult frogs coloured my judgement!' Yet, as things turned out, PPL supplied the cells that made Dolly and put up the cash for that part of the experiment.

So it was that in the winter of 1995–6 Keith, I, Jim McWhir, and Bill Ritchie from Roslin collaborated with Angelica Schnieke and Alex Kind from PPL to clone lambs from three different kinds of cultured cells: from nine-day old embryo cells; from fetal fibroblasts; and from cultures derived from adult mammary gland. In the most important respects the method was the one that we have described for Megan and Morag – we deprived the cultured cells of growth factor for five days to put them into the quiescent, Go state, before fusing with enucleated oocytes (which were supplied as usual by Scottish Blackface ewes). We changed just a few details here and there so as to explore other technical possibilities. In practice, Keith, Bill Ritchie and Karen Mycock did the embryology; Marjorie Ritchie carried out the surgery, assisted by Mike Malcolm-Smith; John Bracken was the anaesthetist; and Douglas McGavin and Harry Bowran injected the sheep.

By the winter of 1995–6 we were, frankly, rather good at cloning embryos from cultured cells; and all three of our cell lines produced live lambs. Some died soon after birth, unfortunately, but no fewer than seven are still living happily at Roslin: four produced from cultured embryo cells; two from cultured fibroblasts; and Dolly, from the cultured mammary cells.

Lambs from embryo cells: Cedric, Cyril, Cecil and Tuppence

PPL supplied the cultured embryo cells – which, like the cells that produced Megan and Morag, came from nine-day-old embryo discs. But, whereas Megan and Morag came from a Welsh Mountain embryo, the 1995–6 culture was from a Poll-Dorset; a pleasant, rounded, lowland breed. These cells were cultured in the presence of leukaemia-inhibiting factor, or LIF, so as to reduce differentiation – in the hope of producing cells comparable with the ES cells of mice. In the end, though, the cultured Poll-Dorset cells were nothing like ES cells, and also quite dissimilar from the cells that produced Megan and Morag. These cells were used as nuclear donors after seven to nine passages.

Bill Ritchie and Keith constructed 385 embryos from these cultured embryo cells, fusing each one with an enucleated Scottish Blackface MII (metaphase II) oocyte. Here they varied the Megan and Morag routine yet again, for they also attempted to raise ninety-two out of the 385 *in vitro* – avoiding the need for temporary recipients. After all, human embryos produced by IVF are raised from zygote to blastocyst in a dish and are then transferred directly to their birth mothers, and scientists seeking to clone livestock would prefer to do the same but so far *in vitro* rearing has not worked so well. In the end, fifteen of the ninety-two embryos raised *in vitro* were transferred into five ewes. One became pregnant, but the pregnancy failed.

But most of the embryos – 231 of them – were transferred in what is now the conventional fashion into the oviducts of temporary recipients. Twenty-seven of these were eventually selected for transfer into surrogate mothers and they yielded four live lambs. These were, and are, Cedric, Cyril, Cecil and Tuppence, the four male Poll-Dorsets we met in chapter 1. The four are genetically identical, yet they are very different in appearance and character.

Lambs from fetal fibroblasts: Taffy and Tweed

The fetal fibroblast cells that were used in the second experiment – the experiment that was exclusively Roslin's initiative – came from 26-day-old Welsh Black fetuses, and they were used as nuclear donors after four to six passages. Tricia Ferrier, the pleasantly harassed mother of one who admits to being a 'gardener of cells', cultured them. From these we made 172 embryos – and again we initially raised some of them in temporary recipients, and some *in vitro*. We transferred thirty-four of the embryos that were raised initially in temporary recipients into ten surrogate mothers, of whom four became pregnant. Two of the four gave birth to live lambs – the two known as Taffy and Tweed. These were cloned from Welsh Black nuclei in Scottish Blackface cytoplasts and by virtue of their Welsh Black genes they are very black indeed. Tricia named them after a Welsh and a Scottish river to acknowledge the breeds that provided their nuclei and cytoplasm respectively, while both share her own initial. We also transferred six of the fetal-cell embryos that were raised to the blastocyst *in vitro* into six ewes, of whom one became pregnant – and she too produced a live lamb. But this lamb died within a few minutes of birth. It is still difficult to raise sheep blastocysts *in vitro*.

Keith maintains that fetal fibroblast cells are probably the best bet for maintaining cell cultures that can be transformed genetically, and then go on to produce live lambs. Polly was made from fetal fibroblast cells. They are robust and grow quickly and can be seen as the functional equivalent of the ES cells of mice – and since they seem to do the job it is perhaps unnecessary to continue trying to produce ES cells in sheep. So, from the point of view of Roslin and PPL, Taffy and Tweed are perhaps the most important clones of all. But for Dolly, they might also have been the most famous.

A lamb from an adult cell: Dolly

The third set of experiments in the winter of 1995–6 involved the mammary gland cells that had been prepared and maintained at Hannah, and came originally from the six-year-old Finn-Dorset ewe. Finn-Dorsets are produced by crossing Finnish rams with Dorset ewes. Finns are very prolific – sometimes they have quins – and Dorsets have a relaxed attitude to seasonality: they can breed twice a year and indeed may be fertile virtually all the year round. When the two breeds are crossed their reproductive qualities are to some extent combined and at one time there was great interest in developing Finn-Dorsets as a commercial cross (though actually, if a ewe has too many lambs she cannot rear them all and artificial rearing raises problems of its own). Anyway, the particular Finn-Dorset ewe who provided the cells was in her last trimester of pregnancy so her mammary cells were preparing for lactation. The cultured mammary cells were used as donors after three to six passages.

Bill and Keith between them constructed 277 embryos from the Finn-Dorset mammary cells. All were transferred into the oviducts of temporary recipients, and 247 were recovered. Twenty-nine of them had successfully developed into morulae or blastocysts. These were transferred into thirteen ewes, of which one became pregnant; and this solitary Scottish Blackface surrogate mother went on to produce a live Finn-Dorset lamb. This was Dolly. John Bracken thought of the name: Dolly Parton was the inspiration, stressing the mammary connection.

What should we make of these results? There was, and is, only one Dolly: only one out of 277 embryos that began with mammary gland nuclei stayed the whole course, and became a live lamb. It remains astonishing that the experiment worked at all – but, even so, this was a skin-of-the-teeth success. If *none* of the 277 had succeeded then sensible biologists everywhere would have been confirmed in their belief that this could not be done. We probably would not have tried again. It certainly would have been hard to get the funding. Others scientists elsewhere would surely have been put off just as they were in the early 1980s when James McGrath and Davor Solter declared that cloning of any kind by nuclear transfer was simply not possible in mammals.

Were we extraordinarily lucky – or were we perhaps unlucky? From a sample of one we cannot produce sensible statistics. Perhaps in another season 277 embryos made in exactly the same way would have produced twenty Dollys. But then again perhaps if we had produced another 10 000 embryos we would have drawn 10 000 blanks. But there are other ways to view the figures: notably, only thirteen ewes became pregnant with the transformed blastocysts, and from this thirteen came Dolly. One success in thirteen is a very good strike rate for a novel technique. It compares favourably with the rate achieved in human IVF.

Neither do we want to claim that we have really 'cracked' the business of cloning. The round of experiments that produced the four young rams, plus Taffy and Tweed and Dolly, also resulted in one neonatal death. There also remains the ever-present spectre of large fetus syndrome – for fetuses of ruminant animals cloned by nuclear transfer are typically about one-third heavier at birth than is normal, and sometimes much more. Gestation is also extended. As we described in chapter 6, the welfare implications of this, and the consequent expense, killed off the cattle-cloning industry in the United States in the 1980s. Our experiments did not seem to fall foul of large fetus syndrome – but we cannot be certain that it did not play a part in some of the pregnancies that failed.

We cannot even be absolutely sure that the lambs that were born alive and healthy were not to some extent affected. This may seem strange, since you might think that all you have to do to see if lambs are heavier than normal is to weigh them. But it can be surprisingly difficult to judge what the 'normal' weight of a lamb actually is. For one thing, even when sheep are born on ordinary farms under everyday circumstances, their birthweight varies enormously. Scottish Blackfaces born by natural means at Roslin have varied from 3.0 to 9.0 kg. Note that lambs at birth are relatively much bigger than human babies are at birth, since young lambs must be active and able to follow the flock more or less from the word go.

All kinds of factors influence the birthweight: whether the mother produces a singleton or twins (or triplets); and whether she gives birth out on the hills, or in some lowland shed. Genes matter too, of course. The genotype (genetic make-up) of the mother of course has an influence, so that Scottish Blackface normally produce bigger lambs than Finn-Dorsets would do. The genotype of the lamb matters, too, so that if a ewe from a small breed is crossed with a ram from a larger breed (as is common on commercial farms) the lamb is bigger than it would be if the ram had been as small as the mother. In our Dolly experiment we created an embryo by transferring genes from a Finn-Dorset into the egg of a Scottish Blackface, and then transferred the embryo into the womb of another Scottish Blackface, and then kept the surrogate mother in cosseted conditions in a shed. What on earth *should* her lamb have weighed at birth? Should the lamb have been within the weight range of a Finn-Dorset, or a Scottish Blackface? Common sense suggests that either would be reasonable; and, as there are no precedents for an animal like Dolly, there are no data by which to make a judgement.

In fact, Dolly weighed 6.6 kg at birth. This is heavy for a Finn-Dorset, at least in our experience at Roslin: Finn-Dorsets born on our farms have weighed from 1.2 to 5.0 kg. But as we have seen, this is well within the normal range of a Scottish Blackface. Some of the other lambs were also on the heavy side at birth, though Dolly was the heaviest; so all were well within the normal range of their Blackface mothers. In short, it doesn't *seem* as if large fetus syndrome was significant in that season, but we cannot affirm that it did not play a part. We continue to investigate large fetus syndrome at Roslin. We would dearly like to know the cause of it – for reasons relevant to human medicine as well as to agriculture and pure science.

Large fetus syndrome is associated with prolonged pregnancy.

As Keith is wont to say of calves, 'They seem reluctant to be born.' Length of gestation in sheep is not determined primarily by the genotype of the mother but by the genotype of the fetus; the fetuses of each breed 'know' how long they need for development and 'tell' the mother when they are ready to be born, through hormonal signals that are not yet fully understood. The gestation period of all the cloned lambs born in 1996 was longer than their breed average. Thus gestation in a Finn-Dorset averages 143 days, while Dolly was born on day 148. Black Welsh lambs are generally born after 147 days, but Taffy and Tweed took 152 and 149 days respectively, and the one that died within a few minutes of birth had taken 156 days. Poll-Dorset gestation averages around 145 days, but the lambs that were derived from embryo cells (Cyril, Cedric, Cecil and Tuppence) took between 148 and 152 days. So – yes, the total gestation does seem to be prolonged, but not dramatically. None of the births were induced. It may simply be that transferred embryos take longer than usual to open their hormonal dialogue with their surrogate mother.

We did not announce the birth of Dolly and the others to the world at large immediately. There were genetic tests to do (of which more later) to ensure that the various sets of lambs were indeed genetic clones of each other, and of the cultured cells that had provided their nuclei. We then had to write the account of what we had done (complicated by the multiplicity of authors, and the collaboration of two separate institutes), then send it to *Nature*, who had to have the paper refereed, and so on. In the end it appeared in *Nature* on 27 February 1997 with the grand title, 'Viable offspring derived from fetal and adult mammalian cells' by I. Wilmut, A. E. Schnieke, J. McWhir, A. J. Kind, and K. H. S. Campbell. Keith was listed as the first author in the Megan and Morag paper; so (as we agreed) it was my turn to be listed first. By convention, the second-best position in the listing of a multi-authored paper is last.

It may be that I had greeted Dolly's birth, about seven months earlier, somewhat too soberly. But when we announced her existence to the world at large, the response was extraordinary. Overwhelming. We anticipated it up to a point: Megan and Morag had attracted plenty of publicity. But we knew there would be even more so we were at pains to prepare. Harry Griffin, Roslin's bearded, affable, much-respected and eminently capable assistant director, worked with PPL and their PR company, de Facto, to ensure that everything ran smoothly. I was appointed to act as the front-man for Roslin and Ron James, as managing director, represented PPL.

But nobody could have anticipated what actually happened. To begin with, the news was leaked ahead of time. *Nature* publishes weekly, on Thursdays. On the Friday before each edition it sends out press releases to tell the principal media, outlining the main stories. But there is an embargo: no one is supposed to report on *Nature*'s contents until the Wednesday evening, the day before the journal's publication. I am told that most journalists respect embargoes most of the time since, in general, it is in everyone's interests to do so; but, it seems, few pay much attention to any press handouts that arrive on a Friday afternoon since they cannot make use of them until the following week. Besides it is said that on Friday afternoons, journalists are often tired. So we planned a press conference for Tuesday: four days after the press handout, and two days before the *Nature* publication. That way, we reckoned, everyone would have a good chance to take stock and get the facts right.

But in the summer of 1996, soon after Dolly was born, a television producer called Christopher Martin was making a film about biotechnology and he happened to be visiting Roslin – mainly to talk about Megan and Morag. His film was not due to be shown until the spring of 1997, after the Dolly paper would be published. So Harry Griffin and I decided that it would be reasonable to tell Martin about Dolly – and indeed that it would be somewhat unfair not to. Martin wanted his film to be shown as soon as possible after the formal Dolly paper was published but, he says, *Nature* would not give him a firm date – which, he said, 'made my life a bit difficult'.

So when *Nature* circulated its own press handout on Dolly, six days before the formal publication, Martin was one of the very few people outside Roslin's and PPL's inner circle who already knew about her. He also wanted to draw attention to his own film, which had finally been scheduled for broadcast a few weeks later. So he told a few people that they should look especially carefully at the press handout from *Nature*, and among them was Robin McKie at London's *Observer*. McKie then decided to break the embargo. The *Observer* comes out on Sundays and he knew that if he waited until the following Sunday to publish, the news would already be old hat. McKie's excuse for jumping the gun was that he found out about the work from Martin – a source other than *Nature*; and if you find out by a back-door route then convention has it that embargoes no longer apply (or so I'm told). Martin insists, however, that he merely drew McKie's attention to the embargoed press handout. In any case McKie published a full-length account of Dolly in the *Observer* four days before the *Nature* publication.

It is to McKie's and his editor's credit, however, that they did see the significance of Dolly. Not everybody did. Martin tells us that when he drew the attention of ITV (Britain's overseeing commercial TV company) to the forthcoming *Nature* paper the editor first asked, 'What's cloning?' and then, when reminded of Megan and Morag, asked 'But didn't we do this last year?' Finally, when Martin explained that a clone of an adult cell was a totally new kettle of fish the editor asked, 'Can you guarantee that this will be world news?' Hot stuff was already lined up for the Thursday evening, said the editor. Aussie wit Clive James was booked to interview champion racing driver Damon Hill – 'and we can't possibly shift that'.

But once the *Observer* had published, everyone else rushed to follow suit, for newspapers follow each other around at least as anxiously as any flock of Finn-Dorsets. The *Sunday Times* and the *Sunday Telegraph* managed to get stories into their late editions, within hours of Robin McKie's piece. The press agencies – Agence France, Associated Press, Reuters – then transmitted the stories worldwide and the American Sunday papers, with the six hour advantage of their time zone, were able to run full accounts; the one in the *New York Times* was three times as long as the *Observer*'s, and contained a great deal more science. Our press conference, which we had cosily planned for the following Tuesday, had to be brought forward to the Monday (that is, to the following day). While Harry Griffin and I spent the Sunday answering telephone calls from around the world at the Institute, Ron James did the same from the de Facto office in the south of England where he was spending the weekend. However, this was only the very beginning, for during the next five weeks many aspects of life were disrupted by the media interest.

Collectively in the weeks that followed we dealt with 2000 telephone calls. At times, the number of calls overwhelmed the receptionists on the switchboard and colleagues complained that they could not make telephone calls because all of the lines were engaged. Next in the line of communication were our secretaries: Jaki Young for me and Frances Frame for Harry bore the brunt of the onslaught. Both worked frantically and patiently to coordinate interviews and to cope with the sometimes overbearing insistence of reporters in demanding that they be put through immediately to a spokesperson. Between us all, we talked to a hundred reporters and also arranged for Dolly's picture to be taken by sixteen film crews and fifty photographers. Even now, more than two years after this initial publicity, we still get a steady flow of such requests.

Dolly, admirable creature that she is, responded splendidly and became even more tame and spoiled than she was already. Harry says he learned a great deal from the experience, not least about the realities of time zones. Thus Dolly was scheduled to lead *Good Morning America* on the Monday morning and Harry was supposed to arrange for me to give an interview in the Roslin car park at noon. He innocently assumed that 'noon' meant roughly lunchtime, as it does for most purposes, only to be reminded very sharply that American television in general and TV satellites in particular wait for no man – even scientists who have just cloned a sheep. The moment passed and Harry's popularity in the US took a definite dive.

Worldwide the press reaction was of all possible hue: sober, thoughtful, pompous, portentous, shrill, frivolous, whimsical, and just plain daft – the full panoply of human response. *Time* homed in with its customary efficiency and showed Dolly, or rather two identical Dollys, on its front cover of 10 March: 'Clone on the range' it trumpeted, and, 'Will there ever be another ewe?' Inside, this became 'We will see ewe again'. Its fourteen-page special report began, memorably, 'One doesn't expect Dr Frankenstein to show up in a wool sweater, baggy parka, soft British accent and the face of a bank clerk. But there in all banal benignity he was: Dr Ian Wilmut.' I imagine that bank clerks would find this extremely irritating. *Der Spiegel* did not exactly sit on the fence. Its cover-lines read 'Wissenschaft auf dem Weg zum geklonten Menschen' – 'Science on the way to cloned people' – followed by 'Der Sundenfall', roughly translating as 'The fall of man'. France's *Recherche* simply asked: 'Clonage: bluff ou revolution?', which I think even I can translate. From the depths of a Chinese newspaper Dolly gazes benignly through thickets of ideograms. News travels.

The more sober headlines ranged from John Lloyd's 'Good or bad, Dolly has a role to play' in *Scotland on Sunday* (the Scottish press made more of Dolly than the English did – a welcome change from Glasgow Rangers and devolution) to Clifford Longley's, 'Alas, the scientist is not to be trusted' in the *Daily Telegraph*. Others let their imaginations run. 'Dolly the cloned sheep kills a lamb – and EATS it!', shrieked Mike Foster in the *Weekly World News*. He quotes 'a frightened researcher' who told him: 'When you do something to anger her, she looks at you with those intense red eyes – eyes full of hate.' In a tireless search for truth Mr Foster then sought the opinions of scientists and religious leaders. An anonymous zoologist assured him that Dolly's behaviour was 'most unusual' while the generously moustachioed but 'worried' Rev. Bob Tweeny of Boston suggested that perhaps, 'A clone simply has no soul'. In reality, Dolly had become the very model of ovine affability; after all the attention she has received, she was, and is, probably the tamest and most trusting sheep in the world.

But most accounts skipped past Dolly as quickly as they decently could and focused on the prospects of human cloning. Predictably, most at first were set dead against it and I and my Roslin colleagues made it very clear that we had no such ambitions. But the opposition was far from universal. Many volunteered to be cloned while others, sometimes very sadly and poignantly, wanted to reconstruct – essentially to resurrect – their deceased relatives. A Canadian cult known as Rael, run by Claude Vorilhon from Montreal, teaches that life on Earth was created by aliens who effected the resurrection of Christ by cloning. Vorilhon accordingly founded an instant company called Clonaid which offered to replicate clients for $200 000. He assured *New Scientist* in May 1997 that 'the technology is not at all dangerous'. He knows more than we do, then. In general, as time passed, the knee-jerk opposition may have faded, so that by December 1997 Professor Lori Andrews of the Chicago-Kent College of Law was telling Gina Kolata in the *International Herald Tribune*, 'I see a total shift in the burden of proof to saying that unless you can prove there is actually going to be harm, then we should allow it.' I and Keith remain firmly against human cloning, however. We will discuss this in more detail in chapter 13.

Steve Connor, then of London's *Sunday Times*, criticised us on welfare grounds for continuing to clone sheep after we had already shown, in principle, that it was possible. Later he criticised us for not producing more than one Dolly. Sometimes you just can't win. Tom Wilkie and Elizabeth Graham of the Wellcome Trust, London, criticised the way in which we made the facts about Dolly known to the world. They suggest that we could have spoken out much earlier, or that there could have been a public debate at the time the experiments were first mooted – for, they said, people at large were presented with a *fait accompli* – and, in particular, that in Britain at least very few scientists apart from the Roslin inner circle commented on the experiments; although in the US many scientists from many disciplines were keen to have their say. In particular, say Wilkie and Graham, Britain's Medical Research Council might have issued some official statement and guidelines. They also suggested that Roslin itself could have presented fuller information when it finally announced Dolly.

Yet we did announce Dolly as quickly and decently as we could – given the need to make sure that she was indeed derived from the cells she was supposed to derive from, and the inevitable time lag before scientific publication. Many of the press reports were very much to the point and the variety of response might be said to represent democracy in action. *Pace* Wilkie and Graham, perhaps the best route is to provide the bare facts and let the comment run and run – as indeed has been the case. Dolly affects the whole sweep of human history and everyone must have their say.

Others, however, criticised the Dolly experiments on scientific grounds. Some asked, 'Is Dolly really a clone of an adult cell? Is she really what Roslin is claiming?' This criticism was largely misguided but it was made in a prominent place by professional scientists, and so demanded a response.

Is Dolly the real thing?

Dolly is of tremendous significance because she is derived from a differentiated, adult cell; yet this bald statement requires two conditional clauses. First – a serious point – the cell that produced Dolly might not be as differentiated as all that; and secondly – though this is a point that Keith and I have always felt was fatuous – two scientists have questioned whether Dolly in fact came from the adult ewe's tissue at all.

We made the first of these points ourselves in our original paper in *Nature*, in February 1995. We stated quite baldly: 'The phenotype of the donor cell is unknown.' Mammary tissue – like the tissues of all the body's organs – contains various types of cell. Ninety per cent of the cells that were cultured to make Dolly were indeed mammary epithelial cells (OME) but there were also 'other differentiated cell types, including myoepithelial cells and fibroblasts'. So, as we said, 'We cannot exclude the possibility that there is a small proportion of relatively undifferentiated stem cells able to support regeneration of the mammary gland during pregnancy.' Many tissues contain such 'stem' cells, which, typically, are pluripotent: able to give rise to all the various cell types of the particular tissue. The liver and blood, for example, are constantly replenished by such stem cells. If Dolly did derive from a stem cell it would take away some of her lustre, for although stem cells are certainly differentiated to an extent they are not necessarily committed totally to only one course. From a practical point of view the distinction may not matter so much: after all, all adult animals contain many stem cells, and if some of these can be coaxed back to totipotency then cloning of adults would indeed be possible, which is the main point. But we still have never claimed to have shown beyond all doubt that a fully differentiated cell from an adult can be restored to totipotency. There is no need to doubt that this is possible, after Megan and Morag, Taffy and Tweed, and Dolly; but Dolly did not demonstrate beyond doubt that this was the case (although other experiments since, in other laboratories, *have* demonstrated this).

Two more objections were raised in a letter to *Science*, the American equivalent of Britain's *Nature*, in February 1998 by Norton Zinder of the Rockefeller University of New York and Vittorio Sgaramella of the University of Calabria in Italy. First, Zinder and Sgaramella pointed out that the old Finn-Dorset ewe who supplied the cell that made Dolly had been in the last trimester of pregnancy at the time of her death – because, as we have seen, PPL and Hannah wanted to study mammary cells that were preparing to produce milk. So the old ewe was carrying an advanced fetus. Occasionally, at least in pregnant women, cells are known to break free from the fetus, and find their way into the mother's bloodstream. So, asked Zinder and Sgaramella, did the mammary tissue of the old ewe contain stray fetal cells? And was Dolly made from one of these? If so, then she is not what we were claiming. Even if Zinder and Sgaramella were right, of course, Dolly would still have been remarkable – since the cell that gave rise to her would have derived from a fetus late in pregnancy, which at the time was the oldest source of donor nuclei ever employed. Taffy and Tweed would have been the only close rivals at the time, and they derived their nuclei from cells from a 26-day-old fetus. Even so, a cell from a third-trimester fetus would not be so impressive as one derived from adult tissue. Secondly, Zinder and Sgaramella complained that there was only one Dolly. Neither we, nor anyone else, had repeated the experiment. And in *Science*, they said, 'one successful attempt is an anecdote, not a result. All kinds of unimagined experimental error can occur.'

Some people might think that these objections were worth making but we were soon able to answer them beyond equivocation. First, we asked molecular biologists in Sir Alec Jeffreys' laboratory in the University of Leicester – an expert and independent group – to carry out genetic fingerprinting. The Leicester studies showed that Dolly's DNA was identical with that in the cultured mammary cells that had given rise to her. This finding was published in *Nature*, on 23 July 1998 – and it seems especially cogent since Alec Jeffreys developed DNA fingerprinting in the first place, in 1984. A second study published in the same issue by Rosgen, PPL and Roslin confirmed that the two DNA samples were indeed identical.

However, genetic fingerprinting can be carried out with various degrees of sensitivity. In addition to genes – stretches of DNA that code for proteins – DNA macromolecules also include small, repeated sections known as 'microsatellites'. Their function is unknown: it is quite possible that they do not contribute to the wellbeing of the creature that contains them at all. They might simply represent bits of DNA that are good at replicating, and do no harm, so do not get edited out. The microsatellites vary enormously: so much, that no two individuals have exactly the same pattern. Nevertheless, different individuals will have *some* microsatellites in common – and the more closely related they are, the more microsatellites they have in common. An individual's uniqueness is revealed only when you start to look at many different sites. Of course, when Dolly was born, PPL carried out microsatellite analysis to ensure that her body cells were indeed identical to those of the original ewe, which were (and are) still in store at the Hannah. As already noted, this is one reason why we did not announce Dolly's birth immediately: we needed to complete this analysis. However, Sgaramella and Zinder argued that the PPL scientists had not initially looked at enough microsatellite sites. Sheep are 'highly inbred', they said, and two related sheep could well have quite a few such sites in common. Thus, they claimed, the initial fingerprinting did not show beyond all doubt that Dolly had come from one of the ewe's own mammary cells. They might still have come from a stray fetal cell. In inbred animals the fetus would be very similar to its mother.

In passing we may note that the Finn-Dorset cells at the Hannah are the *only* preparations from the Finn-Dorset breed in any of the laboratories involved. Furthermore, Dolly clearly is a Finn-Dorset: no show-ring judge would suggest otherwise. She is absolutely nothing like a Welsh Black or a Scottish Blackface or a Poll-Dorset or any of the other breeds that have ever been involved in experiments at PPL and Roslin. We can be sure, then, on the crudest logical grounds, that Dolly does indeed come from the tissues that Colin Wilde originally took from the old Finn-Dorset ewe in 1994.

But what of the more detailed charge, that Dolly comes from a stray fetal cell? Well, in our reply in *Nature* in July 1998, we first pointed out that Britain's commercial sheep are not inbred since farmers are careful to maintain genetic variation. Sheep in a field may look all alike to the casual observer, but the shepherd knows better. Human beings probably look much of a muchness to sheep. The ram who had mated with the original ewe would not have been related to her, so the fetus she was carrying would not be especially similar to her.

Then we pointed out that the evidence that fetal cells may circulate in the maternal blood comes from human studies. But the proportion of circulating fetal cells is low, as might be expected: between one in 100 000 (which in fact seems remarkably high) and one in a billion (which seems more likely). The proportion of such circulating cells in sheep is unknown – but ovine and human placentae differ in structure such that, in sheep, there is much *less* contact between the fetal and the maternal circulation. In sheep, then, fetal cells seem less likely to find their way into their mother's blood than is the case in women. In short, say the scientists, there are likely to be very few fetal cells indeed in the body tissues of a pregnant ewe and 'We think it highly unlikely that the partially purified mammary cells could have been overgrown by contaminating fetal cells during their relatively short time in culture.'

The microsatellite tests which the Roslin (with Rosgen) and PPL scientists carried out before we announced Dolly to the world show that the possibility of any mistake is very small indeed. The molecular biologists looked at five microsatellite sites and estimated that 'the probability that another sheep from the same population would have the same genotype as the six-year-old ewe' would be between one in about 2000 billion and one in 270 billion. In short, we concluded in our letter to *Nature* in July 1998, 'it is extraordinarily unlikely that Dolly was derived from a different Finn-Dorset animal and, therefore, reject [Sgaramella's and Zinder's] hypothesis that "imagined and unimagined experimental error" occurred.'

Scepticism in science is fine and necessary but after a time it becomes silly. The odds against the idea that Dolly does indeed derive from an udder cell run into many millions. As Alan Colman has somewhat wearily remarked, 'The only other possibility is that Roslin, PPL, Hannah, and everyone else who knew about Dolly were involved in some giant conspiracy.' Later he thought of another possibility: that Dolly had come from an alien, since there are nothing like enough sheep on Earth (let alone Finn-Dorsets) to provide a perfect DNA match for Dolly if indeed she was not a clone of the old ewe at the Hannah. (He was of course joking.)

Dolly is, in short, what we claimed in our *Nature* paper: a clone of a cell from the mammary gland of an adult Finn-Dorset ewe. But whether that cell was itself highly differentiated, or was a less committed stem cell, we simply do not know – and it is hard to see how we could ever resolve this issue beyond all doubt. But the central fact remains, and is extraordinary: mammals can be cloned from adult, somatic cells.

Why, though, if this is possible, did we not produce another Dolly? Why, if she is so important, had nobody else repeated the work? Were Sgaramella and Zinder justified after all in suggesting that Dolly is 'just an anecdote'?

A few reasons for not making another Dolly are that funds are limited, priorities continue to unfold, and time marches on. For Roslin and for PPL, the main task is not to clone animals for the sake of it but to clone cells in culture as an aid to genetic engineering: to increase the certainty and precision of genetic transformation, and to extend the range of transformations poss- ible. To this end, Taffy and Tweed – cloned from fetal fibroblasts – were and are of more practical interest than Dolly. And in the season of 1996–7 – the very next season after Dolly and Taffy and Tweed were created – we produced the natural successor to Dolly – namely Polly. Polly is cloned from fetal fibroblasts, like Taffy and Tweed – but unlike them she has been genetically transformed along the way.

Then again, Sgaramella and Zinder published their letter in *Science* just one year after the Dolly paper. It generally takes months to set up cloning experiments in sheep and cattle, which at present must be the species of choice. Whole flocks and herds need to be bought in, cell lines established, entire platoons of scientists and technicians mobilised. Pregnancy is some way down the line. Then in sheep, gestation lasts five months and in cattle, nine months. There simply wasn't time, in one year, for anyone else, starting from scratch, to produce another Dolly.

Yet we soon had news of other laboratories, working on similar lines. In March 1998 Jean-Paul Renard at INRA, France's national agricultural research agency near Paris, made it known that he and his colleagues had cloned a calf; and they could be absolutely certain that their calf had been produced from a fully differentiated cell. Dr Renard announced his work even before it had been formally published, to put an end to the doubts and speculations that Zinder and Sgaramella had raised a year earlier. Specifi- cally, Renard's team implanted thirty-five cows with sixty-one blastocysts that had been made by transferring various kinds of somatic (body) cells into enucleated oocytes, by the method that we had developed at Roslin. The calf that was already born when Renard spoke out – christened 'Marguerite' – was derived from a muscle cell of a 60-day-old fetus. To be sure, a 60-day-old fetus is not an adult: but it does have specialised muscles, and muscle cells are among the most highly differentiated of all. At that time, too, four other cows in Renard's study were past the mid-term of pregnancy. One of them was carrying a calf made from the nucleus of a fetal skin cell (essentially a fetal fibroblast); and another came from the skin cell of a two-week-old calf. The French team had already shown beyond all doubt that the cells that supplied the nuclei were indeed differentiated – not from lurking stem cells – since all of them expressed specific marker molecules of various types. We will never know for certain whether Dolly herself came from a fully differentiated, adult mammary cell or from a pluripotent stem cell. But Renard's work shows beyond any doubt that unequivocally specialised cells *can* give rise to whole clones, so that what Dolly promised is now shown to be a fact.

Even more striking, however, are the studies announced by Ryuzo Yanagimachi and Teru Wakayama of the University of Tokyo on 23 July 1998 in *Nature*. For they have produced no fewer than twenty-two *mice* by cloning adult cells – specifically, 'cumulus cells' from the ovary. This is highly significant for two reasons: adult cells provided the cell nuclei; and success was achieved in the mouse – the animal that had once seemed most difficult. In fact the result is triply significant because mice in general are so much easier to handle than sheep, and Wakayama's and his colleagues' success means that the full scientific potential of cloning technology can now be realised.

We will discuss the Japanese work more fully in chapter 12. In the next chapter we will continue the story at Roslin and PPL: the climax of our work so far. This is the creation of Polly – the cloned sheep who is also genetically transformed.

CHAPTER 11

The Denouement: Polly

Ian Wilmut and Keith Campbell:

Polly, born in 1997, brought this phase of our work to a climax. She is the one who demonstrates that the genetic transformation of animals other than laboratory mice is truly plausible – that the transformation of farm livestock could become routine, and that all animal species might in principle be engineered, if that is what society wants to happen. Polly in effect combines the genetic engineering technology that produced Tracy (though with refinements) with the method that cloned Taffy and Tweed from fetal fibroblasts – she is, in fact, cloned from a fetal fibroblast that had been fitted with a gene for human factor IX. Polly thus shows how 'pharming' might truly become a precision technology – and also suggests that the genetic transformation of animals can now extend far beyond the mere addition of foreign genes. The removal of genes – called gene 'knockout' – now becomes feasible, opening a quite new range of possibilities; and so too does the alteration (controlled mutation) of genes *in situ*. The new age of animal biotechnology began with Megan, Morag and Dolly but Polly confirmed that this was indeed the case. Polly also, incidentally, represents the last collaboration between us at Roslin. At least, we might well work together again; but, after the work that led to Polly, Keith went to work at PPL.

Polly's creation was very much a Roslin/PPL collaboration. Angelica Schnieke of PPL is listed first among the authors of the paper that announces her birth. Then comes Alex Kind of PPL, Bill Ritchie of Roslin, Karen Mycock and Angela Scott of PPL, Marjorie Ritchie of Roslin, Ian Wilmut, Alan Colman of PPL, and Keith Campbell – still at Roslin when the work was done but transferred to PPL by the time of publication. The paper, titled 'Human factor IX transgenic sheep produced by transfer of nuclei from transfected fetal fibroblasts', was published in *Science* on 19 December 1997. PPL was obliged to tell its shareholders about Polly in advance of publication and *Nature*, who had announced Megan and Morag and then Dolly, somewhat sniffily refused to take a paper whose content had already been released.

Just as a reminder, here again is the logic that lies behind Polly. As Ian described in chapter 2, livestock were first transformed genetically in 1985 – by injecting DNA into the pronuclei of fertilised oocytes (zygotes). This was also the method that finally produced Tracy in 1990. But this method brings all kinds of snags. First, less than five per cent of the zygotes actually take up the gene. Secondly, the gene is integrated at random into the receiving chromatin – and, since the position within the genome critically affects the expression, the transgenes may not be expressed at worthwhile levels even if they do become integrated. Thirdly, the new DNA cannot be injected before the pronuclei have formed and swelled to a sufficient size – and by the time they have done that the DNA within them has already begun to replicate. The novel DNA may not then become integrated until after the cell itself has divided, and so will be present in some daughter cells and not others. The animal that then develops may be a mosaic: some of its cells may contain the transgene, but others may not. In mosaics, the transgene may or may not find its way into the germ-line cells (eggs and sperm) so that, even if some of the animal's cells are transformed, it still might not be able to pass on the transgene to the next generation. Thus at every stage of the operation there is inefficiency.

Finally, it is not possible to tell for certain whether a gene has been integrated into the genome before the putatively transformed embryo is transformed into a surrogate – so that a great many embryos are transferred into a great many ewes that may prove, when they are born, not to be transgenic at all, or if they are they express the gene at too low a level, and/or they fail to pass on the gene in their eggs or sperm. To be sure, it is to some extent possible to assess the genetic status of embryos produced by microinjection. One technique is to use the polymerase chain reaction, or PCR, which enables very small amounts of DNA to be multiplied rapidly until there is enough to analyse – for by PCR it is possible to detect the presence of the transgene directly. Alternatively, a marker or 'reporter' gene can be attached alongside the transgene, which will express in the culture itself. But such methods often give false results. For example, marker DNA may show up as present in an apparently transformed young embryo – but may be lying loose within the nucleus, without being integrated into the genome. Such free-lying DNA will not take part in mitosis and will not find its way into all the cells of the developing animal. Overall, then, Tracy herself was a piece of good fortune. Because of her, PPL continues to flourish. But she was a rare success – and if she had not come along when she did then this whole technology might have foundered.

Scientists who work with mice have found a way around these problems – the way provided by embryo stem (ES) cells. ES cells are derived from the inner cell mass of *some* strains of laboratory mouse, and after culture they can be reintroduced into the inner cell mass of other embryos, and then become integrated into any of the tissues of the developing mouse – including its germ cells. ES cells can be genetically transformed in culture; and, because there are millions of cells in the culture, spread out neat and flat on the culture medium, they lend themselves not simply to the introduction of DNA but to a range of other genetic-engineering techniques as well. Genes can be added, knocked out, or mutated. Furthermore – and this at present is the biotechnologist's grail, which John Clark in particular is eagerly pursuing at Roslin – scientists are working on techniques of 'gene targeting': not merely introducing a gene randomly into the receiving chromatin but placing it precisely, in prescribed regions of the chromatin, so that its expression is predictable and under precise control. Because ES cells can be produced from (some) mice, the genetic engineering of mice has been progressing apace; but in all other animals it has reached something of an impasse.

We have experienced this impasse: no animal apart from mice has yielded ES cells. Cells from sheep and cattle can be cultured well enough, but as they are cultured they differentiate. If transferred back into embryos to make chimeras as in mice, cultured cells of sheep and cattle do not integrate themselves into each and every growing tissue. However, even if the ES cell technique was applicable to sheep and cattle, it would not be as helpful as it might seem. It takes two generations to produce a genetically transformed mouse. First the ES cells are grown and transformed, then these cells are put into a mouse embryo to make a chimera, then that chimera grows up and produces transformed eggs and sperm, and then those eggs and sperm combine to produce pure-bred genetically transformed animals. Nevertheless, the whole operation is quick enough in a mouse, which has a gestation period of a mere 21 days and breeds at ten weeks of age, but such a technique applied to farm livestock would be extremely drawn out. It would take two and a half years to make transgenic sheep by the route that produces transgenic mice: two pregnancies with one period of maturation in between. The same technique in cattle would take at least three years. The decades would soon pass.

So the technique of cloning that was presaged by Megan and Morag, and was brought to fruition by Taffy and Tweed, improves upon the ES route. There is no need to go through the chimera stage. Fetal fibroblasts can be cloned directly to produce entire animals. And if those fetal fibroblasts can be genetically transformed *in vitro* then a new generation of genetically transformed offspring can be produced in one step.

This was the background thinking, then, in the summer of 1996, when our molecular biology colleagues at PPL Therapeutics began work on Polly. Three basic ingredients were required: a gene suitable for transfer, a method of making the transfer, and a suitable culture of cells for cloning.

The transferred gene in fact had two components: that is, it was a 'transgene construct'. One component was the piece of DNA that codes for human factor IX – that is, the human *FIX* gene. The other component was the ovine beta-lactoglobulin promoter: that is, the piece of DNA taken from sheep that normally controls the expression of beta-lactoglobulin, a milk protein, in the mammary gland. This whole construct is code-named 'pMIX1'. Our molecular colleagues first showed that pMIX1 worked very well in mice – that it was incorporated into the genome and did indeed achieve high expression. So it would also be suitable for transfer into sheep. In practice the pMIX1 construct was transferred in the company of a 'marker', a piece of DNA whose presence is easily detected after transfer.

The donor cells came from 35-day-old fetuses of Poll-Dorsets – the same hornless, lowland breed that provided the genomes for Cedric, Cecil, Cyril and Tuppence. They were pathogen-free animals (including scrapie-free) as maintained by PPL – suitable for pharming. In fact the cells were Poll-Dorset fetal fibroblasts, and accordingly were code-named 'PDFF'. The team began with PDFF cells from seven fetuses. One (PDFF2) was male, and the rest were female.

We introduced the gene construct and the accompanying marker into the cultured cells with the aid of a reagent known as 'Lipofectamine'. This is fat based, and helps to penetrate the outer membrane of the cells, which is itself fatty. Introducing DNA in this way is known as 'lipofection'. We transformed only some of the cultured PDFF cells by this method. We left others untransfected, to serve as controls. In fact, ten out of twenty-one clones that were transfected were successfully transformed, and proved to contain the pMIX1 construct.

Our work on Megan and Morag, and then on Dolly, all supports the notion that nuclear transfer works best when cells are first deprived of growth factor to put them into Go, and we first had to establish whether genetically transformed cells could also withstand such deprivation. We deprived some of them for five days – the serum in their culture medium reduced from ten per cent to 0.5 per cent. After this an immunofluorescent test, which reveals replicating DNA, showed that none of the cells were in the S (synthesis) phase. If none are in S, then this suggests that cycling has stopped, and if cycling has stopped then the cells are either dead or in Go. In fact, when ten per cent serum was restored to them they carried on growing – showing that, even though they were genetically transformed, they could indeed withstand the necessary period of deprivation.

So now everything was ready. We used four cell types as nuclear donors: two lines of PDFF cells that had *not* been transfected – one male and one female; and two lines of female cells, which were transfected either with more than ten copies of the pMIX1 transgene, or with fewer than five copies.

Then we (mainly Bill Ritchie) constructed embryos from each of the four lots of cells, using enucleated oocytes from Scottish Blackface ewes as cytoplasts, exactly as we have described for Dolly and all the rest. By now, the technique was routine.

We made eighty-two embryos from non-transfected male PDFF cells, which we transferred into temporary recipients. These in turn yielded five embryos, which we transferred into two surrogate mothers, who both proved to be pregnant after 60 days – one with only one fetus, and one with twins. The singleton lamb was born alive but the twins perished; one died, and its death apparently killed the other. One live lamb out of eighty-two original embryos is a 1.22 per cent success rate. One live lamb out of five pregnancies is twenty per cent success.

We made another 224 embryos from non-transfected female PDFF cells; and these yielded twenty-two embryos after transfer into temporary recipients. We placed all twenty-two into nine surrogate mothers of whom four were pregnant at 60 days, and subsequently produced three live lambs. Three out of the original 224 reconstructed embryos represents a 1.34 per cent rate of success.

We constructed a further eighty-nine embryos from cultured PDFF cells that contained more than ten copies of the transgene construct. From these we obtained nineteen embryos that could be transferred into seven surrogate mothers. Four were pregnant at 60 days – two of them with twins, so there were six fetuses in all. But there were only two live births. Two lambs from eighty-nine embryos represents a success rate of 2.25 per cent: the highest success from the four groups.

Finally, we made 112 embryos from transfected cells that contained fewer than five copies of the transgene. Of these, twenty-three developed well in the temporary recipients, and twenty-one were transferred into six surrogates. Only one of these six ewes was pregnant at day sixty, and this animal was live-born. So the success rate – one live lamb out of 112 embryos – was 0.89 per cent.

All in all, then, our efforts in the winter of 1996–7 yielded seven live lambs, including three that carried the introduced gene. One of them expressed the gene at a particularly high level; and she is Polly. As we said in our paper in *Science*, 'This work is the first example of cell-mediated transgenesis in a mammal other than a mouse.' Polly and her contemporaries, in short, began the new age of genetic engineering in animals.

Still, though, we do not claim to be out of the wood. In all cases gestation was prolonged. The breed average for a Poll-Dorset is 145 days; but all except one of the lambs that went to term had to be induced, which was done at day 153. There was also a high death rate in late pregnancy; at day sixty of pregnancy the ewes carried fourteen fetuses between them, but only seven produced live births. The mortality rate was exacerbated by the two twin pregnancies in which the death of one lamb late in pregnancy may have led to the death of both. The mortality rate for non-twin pregnancies was 28.6 per cent: several times higher than the eight per cent experienced with normal breeding though, as we point out in our paper, this is 'similar to that observed after nuclear transfer using embryonic blastomeres'. In short, genetic manipulation *per se* does not seem to have increased the death rate. Overall, as we said in *Science*, 'An increased understanding of the interaction between the transplanted nucleus and the host cytoplasm, the relationship between the early embryo and the maternal environment and improved culture systems will increase the success of embryo production and manipulation *in vitro*.' But, for the present, we were obliged to acknowledge 'the technique is still in the early stages, and problems remain to be addressed'.

But even though this was a pioneer experiment, and even though the mortality was far higher than is desirable, the results already suggest that transgenesis using the nuclear transfer cloning technique is more efficient than transgenesis by pronuclear micro-injection. In their attempts at microinjection between 1989 and 1996, our colleagues at PPL employed a total of 2877 ewes – including oocyte donors and surrogate mothers – to produce fifty-six viable, transgenic lambs. Thus, each transgenic lamb produced by microinjection required input from an average of 51.4 ewes. The 'Polly' experiments, in which cultured PDFF cells were transfected, involved 104 ewes – oocyte donors, temporary recipients, and surrogate mothers – and produced five transgenic offspring. So for each transgenic lamb produced by this route only 20.8 ewes were involved. As we said in *Science*, 'The most significant difference [was that] no recipients are wasted generating non-transgenic lambs.' We can carry out detailed tests of the transfected cells before they are employed as karyoplasts, to ensure their transgenic status.

Then again, when we use fetal cells to begin the cell culture we know their sex in advance, and often the sex matters. If stud animals are needed, then this means males. If animals are needed to produce exotic proteins in their milk, then the order is for females. If the scientists do not know what sex of animal they are starting with – which they do not if they use zygotes as recipients – then fifty per cent of the animals that are successfully transformed will still be useless.

Such advantages are immediate. Much grander possibilities lie in the future. For one thing, simply adding genes to a genome – as human *AAT* (alpha-1-antitrypsin) was added to the genome of Tracy, or human *FIX* to Polly – is only the first step in genetic engineering. For a great many different reasons it is often highly desirable to remove, or 'knock out', a gene. For other purposes it is advantageous to alter an existing gene – or, perhaps, in the fullness of time, to carry out even fancier manoeuvres, perhaps moving genes from one part of the genome to another, to alter their expression. Such manoeuvres cannot be carried out simply by manipulating zygotes, but they can be done in cultured cells, just as they are now being done in bacterial and plant cells, and in the ES cells of mice. Thus the technique that has produced Polly opens up the whole range of genetic engineering within farm animals – and, in principle, to all animals.

Secondly, at present genes are simply dropped randomly into the genome, and attach where chance dictates. But their position within the genome greatly affects their stability (whether they will indeed be passed faithfully from cell to cell as the animal develops, and from generation to generation) and their expression. Ideally, genes would be placed in the precise part of the genome where they would comply most efficiently with the technologists' wishes. Already, PPL are beginning to achieve success in gene targeting in sheep – although, at the time of writing, we cannot give details for commercial reasons.

Put these two possibilities together and we see why Polly really is significant, and why the next few decades will see such a vast transformation of animal biotechnology in all spheres – pure science, agriculture, medicine, and, if we choose, in animal conservation. We will look at future possibilities in the next chapter.

PART IV

The Age of Biological Control

CHAPTER 12

The New Biotechnology

Ian Wilmut and Keith Campbell:

Biological science and biotechnology were racing ahead even before Roslin and PPL began the work that led from Tracy to Polly, and would be doing so even if Roslin and PPL had never existed. Now, though, with cloning firmly within the modern canon, it is as if the science and techniques of biology have been liberated from constraints that once seemed inviolable. We and our descendants must wait and see what the world makes of this liberation – or rather, must try to see that the new power is put only to good and proper use. It would be foolish to underestimate the potential. Tomorrow's biology, swollen with the new techniques and insights that will accrue from the science and technologies of cloning, now promises us a measure of control over life's processes that in practice will seem absolute. It would be dangerous ever to suppose that we can understand all of life's processes exhaustively: this would lead us into the Greek sin of hubris, with all the penalties that follow. Yet our descendants will find themselves with power that seems limited only by their imagination – that, plus the laws of physics and the rules of logic. In practice it will seem as if the power of human beings over nature is absolute.

Prediction is a dangerous game, but it is one we should never stop trying to play. So let us look at what seems feasible in the light of current knowledge. The first and most obvious possibility is to increase the range of species that can be cloned, and the variety of cell types.

The simplest option: just cloning

Mice are the favourite laboratory mammal because they are small and cheap, have short generation times, produce large litters, and have been assiduously bred over many years to provide dozens of distinct and more or less genetically 'pure' strains, so biologists at the start of the 1980s, notably Karl Illmensee and Peter Hoppe, naturally felt that they would be the obvious initial candidates for cloning. However, as we saw in chapter 6, mice proved to be peculiarly difficult – and the reason given (as Davor Solter confirms in an essay in *Nature* on 23 July 1998) is that in mice, genomic activation (the turning on of genes) occurs very early – at the two-cell or even the one-cell stage; and after the genome is activated it needs to be reprogrammed if it is to give rise to all the tissues of a whole animal. Sheep and cattle proved to be much easier to clone from young embryos apparently because, in them, genomic activation occurs later.

But the lesson from Roslin, and in particular from Megan and Morag, then Taffy and Tweed, and then Dolly, is that cloning is possible even when the cells are differentiated (which is a giant step beyond mere activation) provided the donor nuclei are put into a state of quiescence (Go), which evidently makes them amenable to reprogramming. So in 1998 Teru Wakayama and his colleagues at the University of Hawaii, in the team led by Ryuzo Yanagimachi, did manage to clone mice. Indeed they cloned them from *adult* cells. As they freely acknowledged in their *Nature* paper, Wakayama and his colleagues drew heavily on the Roslin approach. They derived the karyoplasts from cumulus cells which surround the egg in the ovary – and at any one time ninety per cent of cumulus cells are in Go or GI. In these experiments there could be no doubt that the cells from which the mice were cloned *were* the highly differentiated cumulus cells; there could be no suggestion, as there was with Dolly, that the cloned animals had arisen from some less differentiated stem cell. In further deference to the Roslin approach, Wakayama and his colleagues used the MAGIC approach: they did not activate the newly constructed embryos for up to six hours after fusion –

meaning that those Go cells were put into a high-MPF environment. By this means the Hawaiian group produced twenty-two healthy female mice, which they then mated. Indeed by the time of the *Nature* paper some of the cloned mice were grandmothers.

Thus at the time of writing – summer 1999 – it seems that cloning by nuclear transfer is possible in principle in *any* mammal. This cannot be known for certain but there is no good reason to suppose otherwise. So what in practice might be done, and why? Well, even without adding any refinements, such as genetic transformation, there are five obvious areas of application. The first is for research – producing even purer laboratory strains; the second is in agriculture and other areas of domestic breeding – replicating elite animals; the third is in animal conservation; the fourth is for multiplying tissues, as opposed to whole individuals, for use in human medicine; and the fifth is human reproductive clinics. We will look at the last of these possibilities in the next chapter. Here we will look briefly at the first four possibilities.

Cloning for the laboratory

Cloning laboratory animals may seem too obvious to be worth comment, but there is more to it than meets the eye. The central aim, of course, is to produce animals for experimentation that are genetically uniform so that when scientists try out any particular drug or training method, or other procedure, they know that any differences they perceive are due to the procedure, and not to genetic differences between the animals. But there are various difficulties. Notably, the traditional way – and up to now the only way – to produce genetically uniform strains of, say, mice is by inbreeding. Closely related individuals are mated, and their offspring are remated, until a population is produced that is all of a muchness.

But, as everyone knows, such inbreeding is dangerous. It is for this reason that various genetic disorders, including porphyria and haemophilia, have bedevilled various royal houses in Europe.

The problem lies with excess *homozygosity*, which can lead to 'inbreeding depression'. Every individual inherits one set of genes from one parent, and another set from the other parent. If the two parents are not closely related then the two sets of genes will differ somewhat: you might, for example, inherit a gene for red hair from your mother, and a gene for dark hair from your father. Then you are said to be *heterozygous* for that particular gene for hair colour. But if you inherited a gene for red hair from both parents, you would be homozygous for that hair colour gene. The trouble begins when one of the genes in a matching pair is a deleterious mutant – like, for example, the one that produces cystic fibrosis. If you inherit a cystic fibrosis mutant from one parent and a normal gene from the other, then you will not suffer from the disease; your heterozygosity saves you. But if you inherit the CF gene from both parents, you will be affected. Of course, only a minority of genes are as harmful as the CF mutant, but the principle applies broadly: and too much homozygosity leads to the general loss of 'fitness' known as 'inbreeding depression'.

So if you produce laboratory animals simply by inbreeding, then you will perforce produce a great deal of homozygosity, which is likely to lead to inbreeding depression. In fact *most* attempts to produce pure-bred strains of laboratory mice have failed. The strains that exist today are the minority that have survived inbreeding – fortunate beasts that happen, by chance, to lack a significant number of genes that are deleterious, so they avoid the kinds of effects we see in cystic fibrosis. We have to conclude, though, that laboratory mice are genetically peculiar – because most animals simply cannot withstand such a high degree of homozygosity. Yet it transpires, too, that inbreeding does not produce quite such uniformity as might be supposed. Sometimes there has proved to be a remarkable amount of genetic variation (implying heterozygosity) in laboratory strains that are supposed to be completely uniform.

On the other hand, it would sometimes be good to work with creatures that are more 'natural': that is, are more heterozygous. Cloning helps here, as well. It not only offers a route to complete genetic uniformity – at least of the nuclear genes – but also makes it possible to produce strains that are uniform but are *not* homozygous. In fact a highly heterozygous wild mouse – or in principle a wild anything – could be cloned to produce as many genetic facsimiles as required. We are so used to thinking that genetic uniformity can be produced only by inbreeding that we tend to assume that uniformity must imply homozygosity. But consider, say, any one variety of domestic potato. Any particular King Edward or Maris Piper or whatever might well be highly heterozygous – but since it is multiplied by cloning (via tubers), all the individual potatoes are genetically similar to all the others, and so the variety as a whole is uniform.

Anyway: the advantages that may accrue from producing genetically identical laboratory animal strains *without* inbreeding could, as the decades pass, prove very helpful indeed.

Replicating the elite

Similar considerations – and more – apply to the cloning of farm livestock. Thus, on the one hand, farmers seek uniformity: they want to know how their animals are liable to perform under particular conditions, when they are liable to mature, and so on; and so, of course, do their markets. On the other hand, farmers also seek optimum performance – where 'optimum' does not necessarily mean 'maximum', although increasingly this is the case. Within, say, dairy cattle there is a huge difference between the yield of the cows producing most milk – commonly called 'elite' animals – and of those producing least. Thus a wild cow produces around 300 gallons of milk in a year to feed her solitary calf, while many modern Friesians produce 2000-plus gallons. Of course, a modern farmer would have a herd of pure-bred Friesians – but even within one 'elite' herd there is commonly a twofold difference between the best milk-producers (2000 gallons) and the average (around 1000 gallons). In general, farmers seek to bring the average up to the level of the best. But such 'improvements' (this being the technical term) take a very long time. The farmer normally improves his herd by impregnating his better cows by artificial insemination (AI) with semen from an elite bull. But only half the calves thus produced will be female, and each of them takes three years from conception to first lactation (a working year for her own gestation, then a year to mature, then another nine months to produce her own calf as a prelude to lactation). In short, raising the standard of a herd even when the farmer has access to the world's best bulls is a slow business. In breeding time, the average animals are commonly considered to be ten years behind the elite.

But there are further complications. Just as animals (and plants) suffer from inbreeding depression when they are too homozygous, so they can experience what Charles Darwin called 'hybrid vigour' when they are 'outbred', and so are highly heterozygous. So farmers of animals, like growers of potatoes, would in general like to combine overall uniformity with individual heterozygosity. In addition, farmers often seek to combine the qualities of different breeds – so that dairy farmers commonly cross Friesian dairy cows with beefy bulls (such as Herefords or Charolais) to produce calves that are good for beef (since only a minority of calves born in a dairy herd are needed as herd replacements). Among sheep farmers, juggling the options between uplands and lowlands, the crossing permutations can be quite bewildering.

In all such instances – where the need is to raise herd quality quickly, and/or to combine uniformity with heterozygosity – cloning has an obvious role. A dairy farmer might improve his herd significantly in ten years by buying in elite sperm; but he might achieve the same improvement in one season by furnishing his cows with ready-made embryos that have been cloned from some elite animal. No wonder the Americans invested so much in this technology in the 1980s. Of course, ultra-high performance does raise special issues of animal welfare – which alone must set limits on what can be done. On the other hand, cloning and embryo transfer could be of particular value in the Third World where the cattle are vital to the economy and where they are often multipurpose (cows might be required to pull carts as well as to provide calves and milk – though they may feed mainly on straw and must withstand tropical heat), and breeding is particularly difficult because of the many contrasting qualities that are required in any one animal. Whether the economic 'incentives' exist to take the new technologies into poor countries is another question.

Cloning for conservation

Cloning, too, could be of immense – perhaps even of critical – value in animal conservation. Many have doubted this. The critics point out, for example, that the task for conservationists is to maintain the maximum possible genetic diversity within each breeding population and point out, rightly, that cloning does not increase diversity. It merely replicates what is there already. But then, this is precisely the point. Conservationists cannot *add* to the range of genes that currently exists. But they must strive to minimise the rate at which genetic diversity is lost. The great enemy is 'genetic drift' – the steady loss of genetic variation, generation by generation. Thus, when animals breed, each parent passes on only *half* of his or her genes to each offspring. If the animal has hundreds or millions of offspring, like a fly or a codfish, then there is a very good chance that each parent will indeed pass on all of its genes – these being spread randomly among the many offspring. But an animal like a rhinoceros or orang utan may have only about half a dozen offspring in a lifetime, so that some of its genes are liable to remain uninherited. If the population of rhinos or orangs is large then any one variant of any one gene is liable to be contained within many different individuals, so between them the breeding animals should pass on all the genes in the total gene pool. But if the breeding population is low – as it is bound to be if the animal is already rare! – then the less common genes may well be contained within only one or a few individuals, and the individual containing the rarest genes may well finish its reproductive life without passing them on. Hence loss of variation, generation by generation.

Conservation biologists attempt to minimise loss by genetic drift by complicated breeding schemes to ensure that each individual that can breed does indeed mate (while avoiding inbreeding); but these schemes are expensive and difficult to organise. Yet it would be technically easy to take tissue samples (biopsies) from representative samples of all the endangered species of mammals that now exist (about 200 at least are priorities), culture them, and then put the cultures in the deep freeze. (If the biopsies were simply frozen without culturing them first they would probably be damaged. Cultures are two-dimensional – one cell-layer thick – while biopsies are three-dimensional blocks of tissue; and it is hard to freeze a block uniformly.) If the samples were well chosen then between them they could contain virtually all of the genes now present in existing species. In fifty years' time, when the technology that produced Dolly is well advanced, and can be extended readily to other species, and when the species that are now endangered are on their beam ends and have lost much of their present variation through genetic drift, cells from those frozen cell cultures could be made into Dolly-style embryos, and future creatures could give birth to offspring as diverse as those of today. Since the present-day breeding schemes are so difficult to run and organise (among other things, they require cooperation among people who tend to be highly individualistic), the Dolly technology could offer the most realistic option for many of our best-loved and ecologically most significant wild creatures.

Of course there have been flashier proposals, too. Some have suggested, for example, that it might be possible to clone mammoths from samples of frozen flesh dug from the Siberian permafrost; and Michael Crichton in *Jurassic Park* envisaged the re-creation of dinosaurs. It might indeed be possible to re-create the mammoth. Mammoths died out only about 10 000 years ago – and there is evidence that some populations, notably on the off-shore Siberian island of Wrangel, survived for much longer than that – and occasionally flesh is found deep frozen and in reasonable condition. If any cells survived they might be cultured. Mammoths were evidently related fairly closely to the living elephants, Asian and African, so if viable cells can be found it should be possible to re-create embryos using oocytes from modern elephants. These might then be brought to term in the womb of an elephant. All this depends on finding viable tissue, after which all the rest is at least plausible. Perhaps the greatest technical snag at present lies in the obstetrics of elephants. No one has yet implanted any kind of embryo in an elephant.

Dinosaur resurrection, however, will surely remain a fantasy. Dinosaurs disappeared 65 million years ago and no dinosaur flesh has ever been found, or is likely to be. Dinosaur DNA has occasionally been reported but this is disputed, and in any case all the samples are highly fragmented. In the fullness of time it surely would be possible to create viable nuclei from complete samples of DNA; but it can never be possible to create entire dinosaur genomes if bits are missing. The feat becomes technically impossible because it would defy the rules of logic. No one can re-create animals from DNA that does not exist, no matter how much knowledge and skill they may possess.

The more modest aim, though – to reduce loss of genetic variation through drift – could begin immediately. Curators of zoos worldwide should be gearing up to take biopsies from all their animals. Many of the genes those creatures contain have already disappeared from the wild, or soon will.

Variations on a theme: cloning tissues

Late in 1998 Britain's Human Fertilisation and Embryology Authority, in a joint report with the Human Genetics Advisory Commission, gave its blessing to the notion that cloning technology could be employed to culture human tissues that could later be used for repair. Cells from a person would be used to create an embryo, as with Dolly; and then cells from the young embryo would be cultured to provide tissue that was genetically identical to the donor. The embryo itself would of course be 'sacrificed'; but the principle is already established that human embryos up to fourteen days (long before they acquire any distinctive nervous tissue) have not yet acquired the status of personhood. Already it is possible to direct the development of cells in culture to some extent, and to induce them to differentiate in prescribed ways. As knowledge advances biologists will surely be able to induce cells in culture to take any form that is required. Eventually, indeed, clinicians could be turning genes on and off effectively at will. At the same time, biologists are developing ways of inducing cultured cells to grow in and around moulds, so as to form facsimiles of human organs. Putting the two kinds of approach together we can envisage that in a few decades it might be possible to re-create entire complex body organs from cells that are genetically identical with whoever is in need of them. This truly would be 'tissue engineering'. In the meantime, more modestly, we can envisage augmenting and repairing tissues in which such sculpturing is less critical; for example to repair skin after burns. Roslin Roslin Bio-Med, Roslin's own new biotechnology company, is on the case. We will just have to watch this space.

These, then, are the kinds of things that might be done simply by replication – cloning *per se* – though without reference so far to cloning of entire human beings. Things start to hot up when cloning is combined with its natural stable-mate, genetic engineering – as indeed was always the goal at Roslin and PPL.

Cloning with genetic transformation: pharming

Polly illustrates why cloning and genetic transformation make natural bedfellows: why, indeed, cloning technology is needed if genetic engineering is ever to be an exact science and a routine technology (whether, indeed, transgenesis is ever truly to justify the title of 'engineering'). Transformation is best brought about when there are many cells to play with and they are spread out in a dish; and after transformation they can be cultured further and assayed to see if they do indeed contain the required genes, before the cells are made into embryos. Generally, a transformed animal would subsequently reproduce by normal sexual means, to produce an entire lineage of normally reproducing animals that happen to contain the novel, required gene. This keeps things simple; and also allows the breeder to perfect the initial efforts of the engineer by conventional crossing and selection. There is no need to clone sheep such as Tracy, for example. Once they are made they can generate entire flocks of descendants that contain the required transgene and express it to varying extents – of which the best can be selected.

The simplest way to transform an animal genetically by gene transfer is simply to add a gene; and here we can envisage two conceptually different possibilities. The first, and simpler one, is to add a gene that is not particularly pleiotropic (affecting more than one character), and which therefore adds a discrete quality to the animal's existing genetic endowment without affecting its life as a whole. The second route – raising more complications, and in general requiring more knowledge of the genome as a whole – is to add a gene that has many different effects, and does change the animal as a whole.

Pharming is an example of the former course. Tracy and her descendants are perfectly normal sheep – except that they produce the therapeutically useful protein, human AAT (alpha-1-antitrypsin), in their milk. Nothing else about them is in any way altered. Polly has of course been transformed by a more elaborate route than Tracy (involving cloning) but the principle is the same. Polly differs from a normal sheep only because she produces human factor IX in her milk.

Pharming is potentially of huge significance, medically and commercially. Of comparable significance, for the same reason – and a leading concern of PPL – is xenotransplantation.

Organs from animals: xenotransplantation

'Xeno' of course means 'foreign'; and xenotransplantation is the transfer of cells, tissues, or organs from one species of animal into another. In particular, there has been huge interest throughout the 1990s in using tissues and organs from pigs to treat human patients. But pigs must be genetically altered before their organs can be safely transplanted into people.

The surgery of organ transplantation is already wonderful – but two huge problems remain. The first is rejection – the body's immune system treats the transplant as the enemy, and attacks it with all vigour; and the second is supply. In 1996 more than 150 000 people worldwide were on waiting lists for organs, including 6500 in the UK – and less than a third of them actually received a transplant. Demand continues to grow at fifteen per cent per year and it will continue to increase as confidence rises and more and more people are considered as reasonable candidates who once would have been rejected on medical grounds, including diabetics and the over-fifty-fives. The supply of young, healthy organs from human beings is limited for all kinds of reasons. But, as *The Lancet* commented on 29 August 1998, 'Xenotransplantation shows great promise of providing a virtually limitless supply of cells, tissues and organs for a variety of therapeutic procedures.'

But the body's response to a xenograft is far more vigorous – and in some ways qualitatively different – from its response to an allograft (an organ taken from another human). The normal response involves white blood cells and antibodies and nowadays can be controlled well enough with drugs (notably cyclosporin A, which was introduced in the 1980s). But xenografts also provoke a 'hyperacute' reaction. Even before the antibodies and white blood cells get to work – indeed within minutes – the xenograft is attacked by a complex of more than twenty different enzymes which are known collectively as 'complement' and which, among other things, clot the blood of the transplanted organ so it quickly dies from lack of oxygen. Modern drugs can contain the standard antibody and white cell attack quite efficiently, but they do not suppress the hyperacute response – which, for xenotransplants, is the main immunological barrier.

The part of the immune system that provides antibodies and white cells can discriminate between the body's own tissues and foreign tissue – 'self and not-self'. Unless things go wrong (in which case there can be a damaging 'autoimmune' response) the immune system simply ignores the body's own cells. But the complement is not so discriminating. The tissues avoid attack by the body's own complement only by producing specific agents known as 'regulators of complement activity'. However, the regulators work only *within* species. The regulators present on the surface of a transplanted pig organ will not suppress the complement of the human host. So the complement is free to show what it can do – which is plenty.

The answer is to produce transgenic pigs that contain genes which provide human 'regulators of complement activity' on the tissue surface. Grafts from such pigs should allay the hyperacute response. In fact a British company called Imutran started producing such pigs in the early 1990s (and Imutran was then taken over by Novartis). In 1995 David White, director of research, announced that the company's scientists had transplanted transgenic pig hearts into monkeys and *none* out of ten transplants had suffered hyperacute rejection. Two of the monkeys survived for more than sixty days with the transgenic xenografts in place – while the previous record had been thirty hours. In fact the xenografts were not required to support the lives of the monkeys – they were introduced *in addition* to the monkeys' own hearts. But they showed the principle: xenografts fitted with complement regulators from the prospective host avoid the hyperacute rejection, and can then be protected from the normal immune response by the same drugs that effectively protect allografts.

The one great practical problem remaining in xenotransplantation is the perceived risk of infection. Pigs have been favoured as potential sources of xenografts partly because their organs are physiologically similar to those of humans, partly because they are of the right size both in infancy and as adults, partly because they are prolific breeders and so can provide a virtually limitless source of organs, and partly because they share few obvious pathogens with humans (although they do share some, including influenza).

But the obvious pathogens present no great problems: it is easy enough to raise 'germ-free' herds of pigs. The real difficulty comes – or might possibly come – from a specific class of agents known as '*retroviruses*'. Retroviruses are RNA viruses: that is, their standard genetic material is provided not by DNA, but by RNA. When they invade a new host they *reverse* the normal process whereby 'DNA makes RNA makes protein'. In fact the invading virus RNA employs an enzyme known as 'reverse transcriptase', which makes a DNA copy of itself. This viral DNA then infiltrates the DNA of the host genome: in fact the invading DNA carries out a natural form of genetic engineering. Once this rogue DNA is inserted in the host it makes further copies of the virus RNA, which then induce the host's own ribosomes to make proteins. These proteins then provide a protective coat for the viral RNA and so produces a complete virus – for these viruses consist only of coated RNA. The viruses then escape from the host cell and invade new cells, perhaps in fresh animals. But fragments of the DNA that was made by the original invading RNA remain in the host genome and – if they find their way into the germ cells – are passed on to the next generation. Retroviruses have clearly been very successful throughout evolutionary history: every vertebrate species contains DNA from thousands of them, although many are in inactive form. HIV, the agent of AIDS, is a retrovirus. So are many of the 'oncoviruses', which cause apparently 'spontaneous' cancer.

Not all retroviruses are dangerous, however. Indeed, retroviruses in transplanted pig organs would not prove dangerous to the human recipients unless a whole string of conditions were fulfilled. To begin with, retroviruses would have to be present in the grafts. If they were present, they would have to escape from the graft into the human tissues. Once in human tissues, they would have to cause disease (for if they did not, then why bother?). Finally, patients whose lives were endangered by organ failure may well feel that it would be worth their while to run the theoretical risk of retrovirus infection – which, after all, would not necessarily harm them at all. But if the retrovirus were to be passed on to future generations, or find some other mode of transmission to other people, then it would potentially be a public health problem, and *no* risk would be acceptable. The question is, then, which if any of these conditions apply?

Well, some of them are known to apply. Pig tissues undoubtedly do contain retroviruses known as 'porcine endogenous retroviruses' – which abbreviates conveniently to 'PERVs'. In practice it would be impossible to breed pigs without them. Furthermore, some of these PERVs *can* invade human tissue under laboratory conditions. They do not transfer easily from species to species – they do so only when the scientists force them – but what can occur in a lab *might* occur in life. However, it is possible now to test human tissue for the presence of invading retroviruses using PCR (the polymerase chain reaction) to multiply the viral nucleic acids until there is enough to analyse; and such tests have not revealed PERVs in diabetic patients who have been fitted with islet cells from pigs (these being the pancreatic cells that secrete insulin), or in patients who have been hooked up temporarily to a pig's liver while awaiting liver transplant from a human donor. So the small amount of current evidence suggests that, although retroviruses in pig tissue might in theory invade human tissue, in practice they do not. Overall, then, *The Lancet* suggests that 'The regulatory climate is moving towards permitting limited clinical trials in the near future' and the betting surely is that xenotransplantation will play a significant part in future medicine.

So where does cloning fit in? Well, Imutran/Novartis transform their pigs in the same way that PPL transformed Tracy: by adding DNA to embryos. It would clearly be better to produce them by the means that created Polly. Not surprisingly, then, Roslin Bio-Med and PPL now have a great interest in developing transformed lines of pigs for xenotransplantation – the obvious complement to the pharming sheep. As we have seen, too, cloning opens a whole new range of possibilities for genetic transfer. In the future, our companies will not simply add human regulator genes to avoid the hyperacute response. The scientists also intend to *subtract* genes from the pig genome, to produce lines that do not produce the principal antigens that provoke the main part of the normal immune response. If and more likely when pig xenografts begin to be used in human medicine, they might in principle be easier to maintain than untransformed allografts.

Disease resistant livestock

Pharming and xenotransplantation employ conventional farm live-stock but they have nothing directly to do with agriculture. But conventional farming could also benefit from various kinds of genes of a non-pleiotropic kind that produce discrete, *ad hoc* alterations. Notably, nothing is more important to a farmer – whether of animals or of plants – than resistance to disease. Some strains of animals are clearly more resistant to some pests and diseases than others, and often such resistance has a simple genetic basis: a single gene can make all the difference. Thus in the future we can expect that genetic engineering for livestock will often be concerned with disease resistance. In rich countries, enhanced resistance should reduce the perceived need to supplement animal feeds with antibiotics, currently employed as a kind of non-specific curb on infection: a pernicious practice that has greatly increased the spread of antibiotic resistance among common pathogens. In poor countries, which often means tropical countries, we can envisage strains of cattle engineered to be resistant to, say, foot-and-mouth disease or blood parasites – provided the world finds means of disseminating such technologies to societies that cannot easily pay for them.

Most genes, however, have more general effects on the lives of animals. Thus the gene for growth hormone of course causes animals to grow faster and/or to produce more milk, and so is a candidate for the genetic transformation of livestock. But growth involves all the body's systems and growth hormone has multifarious effects – so attempts so far to boost growth by adding growth hormone genes have commonly produced deformity: a disaster both of economics and of welfare. In truth, the genome is not simply the sum of the individual genes. The genes interact; the whole genome is more like a recipe, variously influenced by each of its genetic ingredients. Until we understand much more about the interactions of genes, we cannot safely tinker with genes whose effects are not tightly focused on some single, isolable quality.

Genetic engineering with precision: gene targeting and genomics

The efficiency and reliability of genetic engineering as a whole will, of course, be enhanced by gene targeting, now a hot topic of research not least at Roslin. The idea is to ensure that the required gene is introduced into the precise region of the host DNA where it will operate most efficaciously. When genes are simply dropped in at random, as now, they may or may not be expressed, and may or may not persist within the genome. The details of gene-targeting technique need not delay us, but the principle is to attach a length of DNA to the transgene which has a particular affinity for a particular region of the host DNA.

But in the end, the whole future of genetic engineering rests with the currently emerging science of genomics: the attempt to catalogue all the genes that each creature possesses, to work out what each gene does and, ultimately, to understand the interactions between genes. Complete understanding of any one creature's genome – if such is possible – surely lies decades in the future. HUGO – the Human Genome Project – is the current, worldwide attempt to analyse the human genome. Roslin is currently involved in various genomic projects, to analyse the total genomes of the farm animals. All we can reasonably say at present is – watch this space! These are early days. When biologists truly understand how the genome as a whole produces an entire, intact, working animal, and when they know how each added gene may act within the genome, and can place each gene precisely where they want it, then genetic engineering will be powerful indeed. The endeavours we see at present, dramatic and even fantastic though they may seem (for who would have thought even twenty years ago that a cloned sheep might produce human proteins in its milk?) are simply the foothills of the Himalayas.

Yet all of the above possibilities are actual or putative exercises in gene addition. Genetic engineering in cell culture will make it possible to make other kinds of alteration: to remove particular genes, or (in principle) to change them. Then the genetic engineering of livestock will be at least on a par with that of mice – where such changes are achieved in cultured embryo stem (ES) cells, followed by the creation of chimeras.

Beyond gene addition

Transformed animals could also be used in human medicine in a quite different way: to create 'models' of human disease, so that such diseases can be studied more closely and helpfully. Creation of disease models is based on the notion that the so-called 'single-gene' disorders of human beings are often caused by genes that simply fail to do what they are supposed to. Thus the defective gene that leads to cystic fibrosis is supposed to control the flow of chloride ions across cell membranes, but fails to do this adequately. It would be useful to study cystic fibrosis in sheep, which in pertinent respects are similar to humans – but sheep do not possess the appropriate mutant gene. They could, however, if the healthy version of the gene was knocked out. Of course, this raises huge ethical and welfare issues; but then we have to ask whether the saving of human lives justifies the suffering of sheep. If we decide that it does not, then such 'models' should not be produced. If we decide it does, then the possibility is there.

But animal models and the other technical possibilities floated so far merely represent what is already envisaged with present-day science. Far more significant in the long term is the fact that cloning technology offers great opportunites to study the workings of cells and genes, and so to extend present knowledge by leaps and bounds.

Tomorrow's science

We should again look at the relationship between science and technology. In general, technique – actually making things happen – is technology, and science is ideas and explanations. Science is a cerebral activity, an exercise in thinking; but it needs technology because the scientist specifically sets out to think about the real world, the actual Universe, and needs technology in order to engage with the world, and so to see how it really works. Technology is needed to aid the senses and so to observe – where would we be without microscopes and telescopes, voltmeters and seismographs? It is also needed to aid thought – *vide* the slide-rule and the computer. Finally, it is needed because one way that scientists find out how the world works is by experiment: provoking the thing that is being studied in some way, and seeing how it reacts. At a primitive level, children poke animals to see if they move – and if they do this supports the hypothesis that they are alive. Typically they use a stick to do the poking, because they want to be as far away as possible if the animal does prove to be living – and the stick is the investigative technology. When the children grow into scientists their hypotheses become more refined and so do the provocations that enable them to investigate.

Without the technical means of investigation, scientists would be like children – obliged simply to use their eyes and ears, and poke with sticks. On the other hand, unless there is some underlying hypothesis, that such-and-such a system works in such-and-such a way, so that such-and-such a technique *ought* to produce such-and-such an effect, the techniques would have no direction and no point. Without underlying hypotheses, scientists would be like the metaphorical monkeys trying to type the plays of Shakespeare: hacking out letters at random without any underlying reason, or any sense of what they want to achieve. In practice the scientific ideas and the technological methods proceed in dialogue, each edging the other along. A hypothesis suggests a technique – a method of investigation: the method, when carried out, provides new observations which demand more refined hypotheses; and these more refined hypotheses suggest further investigative approaches; and so on.

So now consider the kinds of problems that biologists of all kinds would dearly like to understand. How precisely does the process of differentiation really work? How exactly, molecule by molecule, are the various bits of DNA turned on and off in each cell type as the organism unfolds? How – once differentiation is complete – do the finished, specialist tissues sometimes produce a surge of hormone (if that is their function) and sometimes refrain from doing so? How – the bigger question – is the whole operation organised? Which cells are telling which what to do? Even more to the point – which *genes* within the genome are telling others what to do? To be sure, we have partial answers to many of these questions – progress since the 1960s has been wonderful – but still, we are only in the foothills.

On the larger scale, what is the precise nature of the dialogue between the nucleus and the surrounding cytoplasm? What exactly tells the genes to switch on in the developing embryo? Why do these orders come so early in the mouse (at the two-cell stage or before) and so late in the frog (at around 4000 cells)? How does this dialogue progress as the embryo becomes a free-living creature?

Specifically, how is it possible to reprogramme a genome – as must have happened in the creation of Megan and Morag, and even more spectacularly in the case of Dolly? Are there limits to reprogramming? Did Dolly in fact derive from a pluripotent stem cell? Could it be that pluripotent cells can be reprogrammed while extreme specialists – like nerve cells or muscle cells – cannot? If so, why should this be?

But the universal question for all research biologists is – how can such phenomena be investigated? Scientists can study only what they have the means to study. As Sir Peter Medawar commented, 'Science is the art of the soluble.'

Very obviously, the technique of cloning by nuclear transfer, with all its many components and aspects, provides a superb series of routes by which to investigate all these fundamental issues of biology. How do the nucleus and the cytoplasm talk to each other? Well, what finer means to find out than by transferring nuclei of different types – from cells of differing degrees of differentiation, and of various species – into cytoplasm of different types? What – to address one of the principal issues of genomics – is the action of any particular gene, and to what extent is this action influenced by the other genes in the genome, and by the cellular environment at large? Again, what finer way to investigate this than by transferring genes from one organism to another, so that they can be seen in action in a variety of contexts? And cloning provides the means by which such gene transfer can become a routine procedure.

Are there limits to DNA reprogramming? Well, this might be investigated by the route that Ian pursued in the late 1980s and early 90s: seeking to create embryos with karyoplasts (donor nuclei) that are differentiated to a greater or lesser extent. What is the mechanism of reprogramming? Well, much might be inferred from the experiments already suggested – but this would be abetted by the many techniques long established by molecular biologists for finding out what happens at the molecular level. These techniques are multifarious and complicated (molecular biology has spawned vast libraries of vast manuals) but they have one quality in common: take a system that you half understand and interfere with it by adding some reagent that you know acts in a particular way, for example by knocking out some enzyme or binding to chromatin. From the effects of the reagent the investigator can judge what would have been going on in the absence of the reagent. Little by little the *ad hoc* inferences reshape the tentative hypotheses, and harden them into the theories that form the canon, the state-of-the-art understanding.

Thus the techniques of nuclear transfer, cloning, and the genetic transformations that these techniques facilitate so dramatically will make it possible to investigate and eventually to understand all the principal mechanisms of genes and cells. From such understanding derives control. Control might – if we or our descendants choose – be exercised by altering the genes that underpin the various mechanisms. On the other hand, genes might be manipulated less permanently by pharmacology: drugs. Present-day drugs are already immensely powerful and often very precise, and operate on the body cells at all levels: sometimes acting on the cell membranes, to increase or reduce their permeability; sometimes mimicking hormones, and so enhancing their action; sometimes mimicking their structure but not their function – and so interfering with their action. Increasingly, however, we see a generation of drugs emerging that acts directly upon the DNA, or upon the RNA and the ribosomes that carry out the DNA's instructions. The more that we understand the behaviour of DNA, the more it will be possible to control that behaviour precisely through pharmacology. By drugs we might control development, if we choose: not in the horribly destructive way through which some modern drugs have sometimes produced deformities *in utero*, but constructively: enabling fetuses that otherwise might have been deformed to develop healthily. Less ambitiously we might achieve the ambition described above: to prompt cells *in vitro* to develop into nerve tissue, or kidney tissue, or what you will to help patients whose own bodies are failing. If we understand precisely the signals that prompt cells to differentiate in particular ways, and if we can imitate those signals – why should this not be possible?

Of course, we cannot assume in advance that all the things we might choose to do will, in fact, be possible. Some things that we might dream of might simply break the laws of physics – and these laws cannot be broken. Others might defy the rules of logic. It will never be possible to regenerate kidney tissue from mammalian red blood cells, for example, because mammalian red blood cells lack nuclei altogether, and so are devoid of nuclear DNA. But the way that cloning science and technology has developed since Briggs and King cloned frog eggs in the 1950s suggests that nothing we might care to do need be considered impossible, provided we respect the constraints of physics and logic. This might not have been true. August Weismann's view of differentiation might have proved correct: it *might* have been the case that all differentiation involves the process seen in mammalian red blood cells – the loss of DNA. If it were so, then we could not envisage the kind of technology that produced Dolly, since mammary cells would already have lost many of their genes. But although Weismann is one of history's great biologists he happens, in this instance, to be wrong. Differentiation in general merely requires the shutting down of genes; and what is shut down, so it now seems, can be switched on again. So as things are there is no reason to assume that *anything* we might reasonably envisage doing with living tissues might be possible, for living tissues are proving to be remarkably compliant.

In short, in the wake of the research that has produced Megan and Morag, Dolly, and Polly, we can look forward to an age in which the understanding of life's mechanisms will be virtually total; at least, the principal systems will be understood molecule by molecule. From this total understanding will come – if we choose – total control. Of course the word 'total' is too absolute. There will always be deficiencies and inconsistencies. Biology will never come to an end. But for practical purposes we might as well assume that absolute control will be possible. It is not irresponsible – 'sensationalist' – to suggest this. It is irresponsible to imply the opposite – that our power will always be too limited to worry about. We are entering the age of biological control, and we should gird our moral and political loins accordingly.

One last point. People – even very senior people who ought to know better – misconstrue the time-scale of science. They seem to have no sense of history. Thus when Dolly was first announced one well-known professor of biology was quick to suggest (a) that there was nothing new in this, since 'scientists have been cloning since the 1960s'; and (b) that nothing much would come of it anyway in the foreseeable future. After all, said this professor, 'doctors have been promising gene therapy for at least five years now, and nothing practical has come of it so far'.

Well in the first place (point (a)), the cloning of Dolly is clearly in a quite different conceptual league from the cloning of frogs carried out by John Gurdon in the 1960s – as Dr Gurdon himself has emphasised. He was shocked by Dolly. In the second place (point (b)) five years is a ridiculously short time to take exception to.

To be sure, science proceeds at a bewildering pace. There are *thousands* of scientific journals worldwide and, although most of the papers in most of them contribute very little, a significant minority push understanding ahead by leaps and bounds. The field of cloning – *cum* – genetic transformation has advanced by an age since 1980. But the Universe as a whole and living things in particular are vastly complicated and there is an awful lot to find out; and, as we have seen, knowledge does not come easily. So indeed the pace is great – but it still takes decades, or even centuries, to drink any subject to the depths. Edward Jenner introduced vaccination in the 1790s and vaccination has transformed human history (the elimination of smallpox alone is one of the most significant historical events of all time) and yet, in the 1990s, vaccination research is as lively as ever and some of its problems loom as great. We still do not have effective vaccines for AIDS, or parasite disorders like malaria or bilharzia – and still do not properly understand why such vaccines are not forthcoming. Sir Peter Medawar brought immunology into its modern phase in the 1940s but still, more than half a century later, practical problems of immunology remain unsolved in many different contexts – from vaccination through allergy to transplant rejection. Molecular biology properly began with Crick and Watson's double helix in the early 1950s but genetic engineering – applied molecular biology – is still in its infancy, nearly half a century later.

So how long will it take before animals in general, including pandas and human beings, might be cloned routinely? How long before genes can be transferred from any organism to any other, with predictable results, and kept under perfect control? How long before the 'age of biological control' is really with us? Well, Ian has (perhaps rashly) said that scientists could make 'significant progress' towards human cloning within three years – and has suggested that such cloning might be achieved within fifteen years. This at least, he says, 'Is as good an informed guess as any.' Absolutely not is he advocating this, however; but merely suggesting what others might do. To be sure, such a feat would not in itself represent an age of control: it would perhaps be more of a stunt, altogether too chancy. But in fifty years or less we might well see tissue regeneration or redifferentiation more or less to order. Within the same period it might become possible to clone any species we choose, more or less at will. In 200 years we might see changes wrought by the emerging techniques as profound as those brought about in the past two centuries by vaccination. Some of our great-great-grandchildren will still be around in 200 years. The 23rd century is a family affair.

In 500 years – who knows? The point is that if we are serious – if we are not simply trying to score political points and to underplay what has been achieved so far – then we should think in serious intervals of time. The human species ought to be around for a great deal longer: at least another million years. The next five centuries are parochial by comparison – but they will surely see what might properly be called 'the age of biological control'. The last 500 years have taken us from the Renaissance to what we now consider to be modernity. To our not-so-distant ancestors we will seem as quaint and confused as the Tudors seem to us now. It would not be fair to say that Dolly begins the new biological renaissance, but she should be placed high among the century of events that has brought that renaissance about.

But let us be more specific. The species that interests us most is our own. So let us see what cloning technology and genetic transformation might hold for human beings – not simply in the next five centuries, but in the next few decades.

CHAPTER 13

Cloning People

Ian Wilmut:

We believe that the science and technology that will emerge from our method of cloning, and from the genetic manipulations that are thus made possible, should benefit humankind and the other creatures for whom human beings have become responsible in countless ways. In the last chapter we outlined just a few of those that can already be foreseen – and who knows what might develop in the next few decades or centuries?

But it has been obvious to everyone right from the outset that if sheep can be cloned then so, in principle, can human beings. Both of us see human cloning as a rather ugly diversion: as a medical procedure superfluous, and in general repugnant. But although we have patented certain aspects of the technique that does not give us a legal right to tell the rest of the world how the technology should be deployed. Human cloning is now on the spectrum of future possibilities – and we, more than anyone else, helped to put it there. We wish this were not the case, but there it is and will remain for as long as civilisation lasts.

So what should we do? What should our responsibility be? As scientists, closer to the action than most, we take it to be our duty to make the facts as we see them as clear as possible – for although the facts must not be allowed to determine the ethics ('is' is not 'ought') they clearly do bear upon the moral arguments in many different ways. But also, simply as members of a democratic society, we share every citizen's right to express our personal opinion. We do not suggest that our personal feelings should carry any more weight than anyone else's. But it's clear that people at large do attach weight to our opinions and for this reason alone we should say what we think.

This penultimate chapter, then, has three functions. We want to look at the factual issues that relate to cloning and related technologies. Then we will survey the kinds of ethical arguments that are being brought to bear on the specific issue of human cloning – although many of them apply just as strongly to other areas of biotechnology – and we will try to show how the facts of the case bear upon such arguments. Finally, we want to offer our personal opinions on human cloning – not because we assume any particular moral authority, but at least as pebbles thrown into the pool of discussion. Discussions on human cloning and the technologies that might spring from it will last for the rest of time. People the world over still ponder moral issues that were current at the time of Christ, and indeed long before; and in 2000 years' time, or 5000 or 10 000, our descendants no doubt will still be turning over the ethical implications of human cloning and its sequelae even if, by then, it has long since become routine (although we suspect that it may not).

We might reasonably begin by asking whether cloning really is as outlandish as many commentators suggest. Present-day reproductive technologies are already 'fantastical', to borrow Hans Spemann's excellent term once more. Does cloning really represent a new age, or is it just one more technique among many?

Just another reproductive technology?

Reproduction does not come easily to human beings. Up to one in every eight young couples of reproductive age are estimated to suffer some kind of shortfall in fertility: shortage of sperm, or sperm that lacks motility; lack of eggs; blocked fallopian tubes; lack of womb; incompatibility between sperm and reproductive tract; poor communication between egg and sperm; and so on.

Techniques to enhance childbirth are at least as old as civilisation – probably much older – but, apart from various drugs, tinctures and quack remedies, the first-recorded technology specifically to enhance fertility is only about 200 years old. An Italian priest, Lazzaro Spallanzani, impregnated bitches by artifical insemination (AI) in 1782. In the following decade the great Scottish physician John Hunter used a syringe to impregnate a woman with sperm from her husband, who had a deformed penis. AI was recorded from time to time throughout the 19th century and in 1884 William Pencoast at Jefferson Medical School in Philadelphia inseminated a married woman with sperm from an anonymous donor, described as 'the handsomest student in the class'.

But although AI is technically so simple, it was difficult to organise in the early days because the sperm had to be fresh, so ejaculation had to be timed to the woman's ovulation, and this could be embarrassing and inconvenient for both parties. Cryopreservation – freezing and thawing cells without killing them – was developed by 1953 and made all the difference; in particular it became easy to preserve the donor's anonymity (although nowadays children born by AI have sometimes claimed the right to trace their fathers). As a medical practice, AI for human beings really took off in the 1970s, and by 1978 single and gay women in the United States could legally obtain AI by anonymous donor. Many more must have conceived by AI without the mediation of physicians. In *Re-making Eden* Lee Silver estimates that a million people worldwide might now have been conceived by AI.

The next significant technique, *in vitro* fertilisation or IVF, is a huge leap forward from AI but is a logical extension of it. The first baby to be conceived by IVF – inevitably called a 'test-tube baby' – was Louise Joy Brown who weighed in at 5 lb 12 oz at the Oldham and General District Hospital in the north of England on 25 July 1978. Patrick Steptoe was the obstetrician but the driving force was Robert Edwards, a Cambridge physiologist, who had worked on IVF primarily in mice through the 1960s and began his partnership with Dr Steptoe in 1968. It took them ten years to transfer the technique to humans, and this may seem a long time.

But the first attempt in humans could not simply be speculative: it had to be a serious clinical endeavour with a high expectation of success. If doctors really do press ahead with cloning, then they should surely proceed in the same cautious spirit as Edwards and Steptoe. The bullishness that some enthusiasts display is surely irresponsible.

Many different conditions can be helped by IVF. A woman with damaged fallopian tubes might still provide an egg that might be fertilised *in vitro*, and the embryo can then be reimplanted into her womb. If a woman is unable to produce her own eggs then her partner's sperm might fertilise a donor's egg *in vitro* and, again, the embryo could be returned to the woman. If a woman does produce eggs but her womb is damaged, then the couple might produce an embryo *in vitro* to be brought to term in a surrogate mother. Thus by 1994 thirty-eight countries were running IVF programmes with more than a hundred births each, including Britain, where it all began, plus the USA, Australia, France, Belgium, Holland, Malaysia, Pakistan, Thailand, Egypt, Venezuela and Turkey. Worldwide in 1995 there were an estimated 150 000 IVF babies and Lee Silver estimates that by AD 2005, half a million babies might be born by IVF in the USA alone, with many millions more in the world at large. In short, IVF is becoming commonplace. We are already at the point where everyone knows someone who has had a child by IVF, or knows someone who knows someone.

Cryopreservation now assists reproduction in many ways. Cells of all kinds are generally preserved in liquid nitrogen at −196 degrees centigrade; but the trick is to take them down to this egregious temperature, where flesh becomes hard as stone, and bring them back again to normality, without turning the water into crystals of ice that slice the fabric of the cell to pieces. The first cells were frozen and recovered successfully in the 1940s; bull semen followed in 1950, and was to have enormous consequences for the world's cattle industry; and human sperm joined the ranks in 1953. The first embryos – of mice – were successfully frozen and thawed in 1971 and the technique was soon applied to the embryos of rabbits, sheep, goats and cattle. As you will recall, I made Frostie in the late 1970s when I was at Cambridge: the first calf ever to be produced from an embryo that had been frozen, thawed and nurtured in the womb of a surrogate mother. The first human child to be born from a frozen embryo was Zoe Leyland in Melbourne, Australia, on 28 March 1984. Since then, the cryopreservation of human embryos has become routine.

Artificial insemination, *in vitro* fertilisation and cryopreservation may now be seen as standard reproductive technologies. The first may be done with kitchen equipment but the second two are definitely high-tech – scions of science – although they can now be carried out by technicians with the aid of standard kits, albeit directed by clinicians. To this basic repertoire, however, is now appended a steadily growing catalogue of refinements, usually known by acronyms. ICSI, or 'intra-cytoplasmic sperm injection', was developed in 1992 and is now widespread in US clinics. Up to eighty per cent of eggs that are fertilised by ICSI develop normally into two-cell embryos which, once formed, develop to term as frequently as those produced by traditional IVF. ROSNI is 'round spermatic nucleus injection'. In some men, the sperm fail to mature properly; they get to the spermatid stage, when they are haploid and more or less devoid of cytoplasm, but they fail to develop tails and so remain immobile, and cannot penetrate the egg. So they are injected into the egg in tailless form – while they are still 'round'.

The refinements grow ever more extraordinary. Ralph Brinster of the University of Pennsylvania Medical School has now taken early male germ cells that are still diploid – spermatogonia – and introduced them into the testes of other males whose own germ cells have been stripped out (which can be achieved with the aid of drugs). The spermatogonia then mature and undergo meiosis in the 'foster testes', and are ejaculated as mature sperm. Remarkably, spermatogonia can be successfully transferred between species – mouse into rat, for example. Exactly what species combinations are possible is not yet established – particular genetic barriers between species will probably count for as much as overall genetic distance – but it might in principle be possible to mature human spermatogonia in the testes of a pig.

Techniques involving donation of eggs are far newer than those of AI, partly because oocytes are so much harder to collect, and partly because there was no point in collecting them until IVF came on line after 1978. The first successful pregnancy initiated in humans with a donated egg was reported in November 1983 from Monash University at Melbourne, Australia. Egg donation is not pleasant – the donor must undergo a course of hormones which can have uncomfortable side-effects and subject herself to minor surgery – but women do volunteer nevertheless, either in a spirit of altruism to help relatives or friends or in return for money. Perhaps, though, the traumas of egg donation are numbered, for scientists both in Britain and in Sweden are developing techniques to take immature eggs from the ovary by needle biopsy, and mature them *in vitro*. This would provide a virtually inexhaustible supply of eggs. Eggs are difficult to freeze and thaw successfully, but this too can now be done, and in 1998 Britain allowed embryos to be created from frozen eggs.

Many women are able to donate their own eggs, yet need to borrow the womb of another. Surrogacy is an ancient practice, at least in principle: the Old Testament tells us that Rachel, Sarah and Leah, all of whom were 'barren' as the Bible has it, persuaded their husbands to impregnate their handmaidens, and claimed the babies at birth. Such *ad hoc* arrangements must have been common in the past. As a formal clinical treatment – perhaps abetted by AI or embryo transfer – surrogacy has been available since the 1970s, the surrogate mothers again being friends or relatives acting altruistically or employed through agencies.

The three most significant technologies listed above – AI, IVF and cryopreservation – all caused tremendous rows and upsets, which still come to the surface. Perhaps it is right that they do so: human reproduction is a serious business, and whatever we take seriously should be argued through at length and is bound to generate heat. Artificial insemination provoked tremendous opprobrium even when it was first developed for cattle in the 1930s, by Sir John Hammond at Cambridge. Bishops objected to its unnaturalness, while others felt that animals should not be deprived of their sexual rights. In human beings, AI was used more and more in the early decades of this century but far less than it might have been, largely because husbands objected to their wives' insemination; and many men (and women) are still not comfortable with the procedure. In the US, Canada and Britain, AI was held to be adulterous and the children it produced were officially illegitimate – a blimpish piece of law making that persisted until the 1960s.

In vitro fertilisation provoked at least as much opprobrium in the late 1970s and 80s as cloning does now – perhaps more, because people in those days were less attuned to high-tech fertility treatments. The inevitable wastage of conceptuses was one of many objections, seen by many Catholics as a form of abortion, and quite unacceptable.

Cryopreservation raises practical problems. IVF follows super-ovulation, so it is generally possible to make many more embryos at any one time than can immediately be implanted and those that are not needed immediately can be frozen and perhaps used for later pregnancies – saving the trauma and expense of later superovulations. But current British law says that frozen embryos should not be stored for more than five years without the express request of the genetic parents and in the early 1990s the problem of disposal again prompted waves of revulsion and protest. Some saw it as a kind of mass abortion.

Maternal surrogacy remains highly controversial. Surrogate mothers have sometimes changed their minds and wanted to keep the babies they have carried. Critics have suggested, too, that surrogacy is immoral with overtones of prostitution, adultery, and/or incest, or that it is 'dehumanising'. Thus Herbert Krimmel, a professor of law, is quoted saying that 'Surrogate mother arrangements . . . encourage and tempt the adopting parents to view children as items of manufacture'. 'Commodify' has become a common verb in this kind of context. But some ethicists argue that women have a right to be surrogates if they choose, with Professor Lori Andrews of the Chicago-Kent College of Law maintaining that a woman's individual rights should override state interventions – which, of course, is a common theme in US politics.

What can we learn from all these apparent precedents? First, it is clear that people in general (including theologians, lawyers and politicians) tend to object to any new exotic technology that affects the human body, as a matter of course. Novelty in general and invasive medical procedures in particular are innately suspect. As time passes, however, the fuss dies down and eventually – at least sometimes – the technology that once seemed outlandish, even diabolical, is widely accepted as normal practice. AI and IVF are the key examples. With all such procedures, however, a hard core of objectors remains. We may be reasonably sure that in 100 years, or 500, some people will still be objecting to IVF for one reason or another.

If we acknowledge – as the present authors do – that some at least of these reproductive technologies are indeed acceptable, then we may regard the initial distaste simply as a passing phase, which most people will soon get over. Of course people are suspicious at first – and quite rightly; but once they grow used to the new technology, and see what benefits it can bring, their misgivings fall away.

But that is not the end of the matter. It does not follow, just because we grow used to some phenomenon or other, that that phenomenon is good. We grow used to bad things, as well as to good. You don't have to be in Bombay for very long before you begin to ignore the beggars. But the underlying reality remains horrifying; it's just that the visitor is hardened. The public seems largely to have grown used to IVF and accepts it as standard medical procedure – but perhaps people are wrong to be so complaisant. Perhaps they have simply grown accustomed to something that is bad. But to us, at least – now it seems to most people at large, now that the dust has settled – IVF does not *feel* like an evil procedure. In the hands of good clinicians and responsible patients it really does seem to emerge as a benign and worthwhile technology.

So does the same kind of principle apply to cloning? Is the present aversion that many people feel – including us – simply a fear of novelty? Or is there really a qualitative difference between the technology of cloning and the reproductive technologies that are already commonplace, and seem on the whole to be acceptable, even when they seem exotic?

Well of course there is a qualitative difference, which has nothing to do with the nature of the technique. All the current technologies are designed to abet sexual reproduction: the creation of a new embryo by the fusion of egg and sperm. This is the normal method of reproduction among human beings, and indeed among all mammals. Furthermore, the sexual fusion of egg and sperm produces a new individual who is genetically unique. Biologically this fusion is first and foremost an exercise in genetic recombination. Sex ensures that no two human beings – indeed no two creatures of any kind – are ever exactly alike. Genetically identical human beings are produced only occasionally, by the division of an embryo in the womb, at various stages of early development. Among people, as among all mammals and most vertebrates, genetic uniqueness is the rule.

But cloning is not intended to abet normal sexual reproduction. It is exclusively an exercise in replication. To be sure, the creation of a clone like Dolly involves many components of sexual reproduction. Once the embryo is made, it has to be inserted into the womb of a surrogate mother to be brought to term. But the essential part of the creation process – the fusion of eggs and sperm – is missing. Fertilisation is replaced by another process: that of nuclear transfer. Thus cloning provides a *substitute* for normal sexual reproduction. It is a qualitatively new addition to the canon of mammalian capabilities. Crucially, it does not produce a genetically unique individual. It produces a new individual who is genetically identical to some pre-existing individual.

Thus cloning is different conceptually from all previous technologies. In truth, it is also more complex technically than most existing reproductive techniques – indeed it combines elements from several different standard technologies – but technical complexity is not the prime issue. What really matters is the intention, which is to bypass normal sexual reproduction; and the result, which is to create a new individual that is not genetically unique.

This last issue – the sacrifice of uniqueness – is the one that has bothered most critics. We should ask what it really entails. To begin with, are cloned individuals really identical one to another, or are they merely similar? And how similar are they?

Are clones identical?

Sometimes in the womb an embryo splits, and each half develops into a new individual. Then we have identical twins.

Physically speaking, analysed molecule by molecule, identical twins produced in this natural way really do seem to be as identical as it is possible to be. At least, each set of genes, in each new individual, is a facsimile of the other's genes; and, although the cytoplasm is not duplicated precisely, at least in the first instance, the two sets of cytoplasm are at least *qualitatively* the same.

Sometimes the similarities between twins who are created in this way are startling; there are classic cases of twins separated at birth who, when compared many years later, are found to be wearing identical clothes, have the same hobbies, are married to remarkably similar partners, and so on. There are identical twins brought up together who think so alike they can finish each other's sentences. Many twins, however, are irritated by these 'classic' examples, which they regard merely as clichés. They prefer to emphasise their differences; and the contrasts – given the underlying genetic resemblance – can be striking too. There are various reasons why this should be so.

First, genes are not as constant as we imagine. They mutate. Everyone's body probably contains many hundreds of mutated genes. A conscientious analyst who trawled all the body cells (and gametes) of two identical human twins would undoubtedly find increasing genetic differences as they grew older, caused by these constant mutations. This may be a small point in practice since most genetic mutations have little or no discernible effect on the phenotype, but it is a point nevertheless. Nature might intend to produce perfect genetic replication each time a cell divides, but absolute perfection is a tall order.

More important is the point emphasised throughout this book: that genes operate in constant dialogue with their surroundings. The DNA is surrounded by the environment of the nucleus; which in turn is surrounded by the cytoplasm of the cell; which is in touch with the rest of the cells of the organism; which is assailed by the world at large – first the womb, and then the great outdoors. The surroundings affect gene expression. For simple reasons of physics, no two entities can be in exactly the same place at the same time. Identical twins may share a womb; but they do not share the same part of the womb – and there is evidence in many contexts and various species that the difference in location can affect development. The two twin babies may share a cot; but one will be on one side, and one on the other, and their view and experience of the world will be ever-so-slightly different.

Ever so slightly, perhaps; but perhaps enough to make a difference to gene expression. Indeed Davor Solter, who has featured throughout this book, has suggested in *Nature* (in the article of 1998 that followed the birth of the cloned mice in Hawaii) that although two cloned creatures may have identical genomes they may nevertheless *express* different genes; and, so long as they express different genes, they are *functionally* different genetically. Furthermore, the genetic differences are liable to increase as the cloned individuals develop, since a gene that comes on line at one point in time (or fails to come on line) will affect the expression of others that are due to come on line later.

For two reasons then – recurrent mutation and variable expression – even two *identical twins*, formed by splitting an embryo, may be genetically different, physically and perhaps functionally. The differences are liable to be slight, but they can be obvious nevertheless.

Then there is the point that Jaques encapsulates so neatly in Shakespeare's *As You Like It*: that as people grow older (Jaques specifically spoke of men, but the principle applies to women just as well) so they change, radically – from mewling infant in Jaques' jaundiced view of life to foul-mouthed soldier and then to enfeebled 'pantaloon'. How does this happen? A geneticist would say, because each of us, as time passes, runs through our genetic programme. In infancy, the genes appropriate to babyhood are expressed, producing simple, touching souls with big domed heads and tiny hands. In manhood the appropriate genes generate testosterone by the dram so that the beard flourishes, the voice booms, and the temper grows worse. In old age the genomic programme is increasingly in disarray until, as Jaques dyspeptically put the matter, we are *sans* teeth, *sans* everything.

Consider, then, a child who has been cloned from his father or from her mother. Clone parent and clone child are at different points of their genetic programme. Leaving aside differences in mutation and expression, they are operating different permutations of their genes, taken from different points on the programme. Functionally, then, at any one point in time, they will be very different people; just as adults reading this book are different in appearance and attitudes from the children they once were, while children can expect to become different as they grow older. On the other hand, the influence of the genes is pervasive. The cloned child would be working through the same genetic programme that its clone parent was still embarked upon, and could expect at least to some extent to re-enact and re-create the person their parent had been. Clone parent and clone child alike would be constantly confronted by the immediate differences between them, yet also by the overall similarity.

But of course – and biologists are at least as aware of this as anybody – genes do not *determine* in tight detail how a creature turns out. In general, the genome merely sets broad limits on the possibilities. A person may decide to be a slob – or have little opportunity to be anything else – and become fat and short winded; or he or she may decide to train, and might perhaps have been raised in the shadow of a gymnasium, and become an Olympic athlete. Same person, same genes – but completely different outcome. Nutrition in infancy influences height (and hence weight) at maturity. The nutrition of the mother influences the birthweight of the infant, which in turn influences that child's final height and weight. Hence children who are well nourished are taller than their parents who were less well fed; and this effect continues through several generations. In Britain, today's 15 year olds typically tower over their parents. Modern pin-ups, the supermodels, are commonly around six feet tall while the Victorian factory girls who posed for pocket money rarely reached five feet. The effect of environment on personality and on educational achievement is clearly stupendous. This issue has become politically fraught but the generalisation is undeniable. Children who are given well-directed musical education at a young age, for example, may become virtuosi, while those who do not handle a musical instrument until they are in their teens are most unlikely ever to achieve more than competence. Children of diplomats or immigrants commonly grow up to be bilingual, or even polylingual; while those plucked from their homelands and brought up on plantations among people from many lands who share no common language may not progress beyond their own reinvented pidgin. Genes, in short, merely propose possibilities. It is the environment that shapes the final outcome.

We can see the physical realities of this principle within the brain. Thoughts and memories seem to be contained within and framed by the brain cells – neurones – of which there are billions. But what matters most is not the number nor the distribution of the cells, but the connections between them: called the synapses. Two individuals may possess similar neurones, but the neurones in the two individuals become wired up to each other in different ways, and so form different neural 'circuits'. Thus two brains with similar cells are nevertheless qualitatively different. If there are billions of different brain cells, then there is an infinity of possible ways in which the circuits could form. Do the genes determine the circuitry in fine detail? Well, if we have only 100 000 genes, and billions upon billions of possible arrangements of synapses, how could it be so? It seems rather that the genes lay down the broad structure, but the structure then works itself out in ways that are beyond their fine control. The final distribution of the synapses depends, again, on environmental influences, in the womb and out of it. There is no reason to suppose that the elusive quality of 'free will' would be less apparent in a person produced by nuclear cloning than in the rest of us. Once the brain is formed it does its own thing.

All in all, then, the genes lay down the ground rules – but in the end our upbringing and experience make us what we are. Pygmies could never star at basketball, Masai warriors could never be jockeys, and very few of us indeed could ever have emulated Mozart, no matter how many hours we practised. Our genes provide the clay or the marble from which we are made and this of course influences the kind of people we can become; but our own experiences, and our perception of those experiences, imposes the sculpting. Here, then, is another crucial set of reasons why two individuals will be different even if – like 'identical' twins – they arise from two apparently identical cells of the same embryo.

But with clones like Dolly, or Megan and Morag, there is yet another layer of difference. For such clones do not arise in the manner of identical twins. They are not made by splitting apart two cells from an embryo. They are made by introducing a nucleus into the cytoplasm of an egg *which usually is taken from another individual*. Thus Dolly had the same genes as the old Finn-Dorset ewe who provided the mammary gland cells in the original culture, but her cytoplasm came from a Scottish Blackface ewe. The cytoplasm, as we have seen throughout this book in many contexts, profoundly affects the expression of the genes. The cytoplasm even contains genes of its own, within the mitochondria. So Dolly and her clone mother are *not* like identical twins. Dolly is merely a 'DNA clone' or a 'genetic clone' or a 'genomic clone' of her clone mother. We have already listed a series of reasons why true identical twins, produced in the womb by the schism of an embryo, nevertheless grow into different people. How much greater will the difference be if the clones are merely genomic clones, who begin with different cytoplasm?

Since these are early days, and so far there are very few genomic clones in the world, we cannot yet answer this question. Already there are signs, however, that genomic clones may vary significantly. We have mentioned Cedric, Cyril, Cecil and Tuppence, the four young rams whom we cloned from cultured embryo cells at the same time as Dolly. They are genomic clones – and yet they are very different in size and personality, with Tuppence the shrinking violet among the four. People who are now hoping to replicate their beloved cats and dogs by genomic cloning will be disappointed, at least sometimes. The personality of the new-found pet may well differ profoundly from the original; and even the coat colour may be different since the expression of the genes that 'determine' colour is very sensitive to early stimuli in the womb.

In short, cloning of course can and generally does produce sets of individuals who can be almost indistinguishable and, since we are accustomed to seeing human beings (and most other animals) in ones, the similarities can sometimes seem almost uncanny. Yet cloned creatures can be noticeably different, and sometimes remarkably so; and genomic clones differ more, one from another, than conventional identical twins. So those who contemplate human cloning face a paradox. Sometimes the differences between two cloned individuals might be upsetting; but at other times it is the similarity that could give cause for disquiet. We are opposed to human cloning in either case; but, for the sake of completeness, we should look at the various contexts in which such cloning might be carried out, and see where the two different kinds of objection might apply.

Why clone people?

In January 1998 Dr Richard Seed, who was educated at Harvard as a physicist (not a physician!), launched a campaign to clone a human being. He lacked funds, a suitable laboratory, and appropriate expertise and, although his grand plan was bruited on many a front page, Alexander Morgan Capron, a professor of law and medicine at the University of Southern California and a member of the National Bioethics Advisory Commission, remarked somewhat archly that 'it belongs in the entertainment section'. At the time of writing, however – well into 1999 – Dr Seed is still promising to clone a human being by the end of the century. At one point he was intending to begin by cloning himself but later, lest he be accused of egoism, he decided to clone his wife instead.

Unfortunately Dr Seed is not alone in his enthusiasm. Though most commentators have been repelled by the notion of human cloning a significant few have welcomed it – and basically for four kinds of reason. Some would like simply to copy themselves. Others seek to re-create departed loved ones. Some suggest that we should clone outstanding individuals – Mozart, Einstein, Gandhi, Michael Jordan and Marilyn Monroe have been among the nominees. Most commonly, some suggest that cloning simply be regarded as another of the many technologies that already help infertile couples (or indeed individuals – or conceivably consortia) to produce their own genetic offspring. Of course, none of these proposals is technically feasible at present and it may never be possible to make another Mozart even if his body still exists, since his DNA is probably long since decayed (and to copy what no longer exists would be to defy the rules of logic). But a great deal will be possible in the fullness of time, and we should look at these different motivations one at a time.

Why, first of all, should people want to clone themselves? Not everyone has given reasons but Richard Dawkins, expressing such a desire in his essay 'What's wrong with cloning?' in *Clones and Clones* says unequivocally, 'My feeling is founded on pure curiosity'. Others, no doubt, are motivated by a desire to increase their own influence. It has often been suggested that dictators might want to clone themselves – as Ira Levin anticipated in *Boys from Brazil*. Actually we know of no living despot who has expressed a desire to be cloned and perhaps dictators prefer to be as unique as possible (for there are degrees of uniqueness, despite what the grammarians say). Perhaps some people who want to be cloned are simply vain. Others clearly feel that by duplicating themselves they will achieve a kind of immortality – though this as we will see is one of the many misconceptions surrounding cloning.

It is pertinent to ask whether curiosity, vanity, the wish for personal power, or an undoubtedly misguided desire for immortality really are good enough reasons for bringing a child into the world. One of us (Ian) has two daughters and an adopted son, while Keith has two daughters, and we both feel that to bring a new human being into the world is a sacred thing – the most serious thing that most of us do in the course of our lives. The birth of a child affects its parents, its family, society as a whole, and of course, above all, the child. Is any motive apart from an unassuageable desire to be a parent, with all that that connotes, truly acceptable? In truth, even without reproductive high technology, children have often been brought into this world for dubious reasons: to prop up failing marriages, to give status in societies where motherhood is a woman's only route to status, to sell into slavery, and so on.

But that is not the point. We should not add insouciantly to the catalogue of unacceptability.

I have been the main spokesman for the cloning work at Roslin and I have often been asked – although I dread this request – whether it is possible to re-create some dead loved one. I was first asked this over the telephone within days of announcing Dolly's birth, by distraught parents from another country. Over the telephone, or in a lecture hall, it is not possible to say much more than 'No: at this time it is not possible'. The more expanded answer, however, is that there are two kinds of caveats – which, as we have already outlined, are opposite in nature.

First, for all the reasons described above, the child (or cat or dog) that was cloned by the kind of technique that produced Dolly would not simply be a facsimile of the original. We might rejoice in the difference – that the new person or animal was a new individual after all, despite the manner of his or her creation. On the other hand, people who *wanted* a precise facsimile would surely be disappointed. The new baby would not simply follow in its brother's or sister's footsteps. The anxious parents would surely be puzzled and perhaps upset by the difference. Cloned cats or dogs might look completely different from their clone siblings or clone parents; notably, they might have significantly different markings. A puppy cloned from a black-and-white dog would itself be black and white; but the patch over the eye that made the original so lovable might be lacking. Those who are planning to offer a service in cloning pets (and there are already some) had better be sure they don't get sued. You might well find a closer match to your much loved original at the petshop. Dolly's clone mother is not around for comparison but Cedric, Cyril, Cecil and Tuppence already show how different genomic clones may be.

In short, parents or pet-owners seeking a facsimile of lost loved ones could well be disappointed – and this brings me to what I think is perhaps the most important of all the objections to human cloning, and the one I have stressed in many a lecture and article. For what we should surely consider above all – even above the interests of the parents – is the welfare, including the psychological welfare, of the unborn child. Parents may already require too much of their children, at least in expecting them to conform to their own ideals. How much worse would it be if the child had been brought into this world as a replica of some other individual! How would the parents avoid making comparisons with the dead brother or sister? How would this affect the growing child?

Or suppose that the child was made in the image of one of the parents. As I have said to many an audience on both sides of the Atlantic, suppose my wife and I had decided to clone me, and the Wilmut clone was now a teenage boy. My wife and I met while we were both still at school – not the same school, but neighbouring schools – so this hypothetical lad would now be as old as I was when we met. How would Vivienne regard this near-facsimile of the boy she once fell in love with? How would he think of his own future as he regarded me – now ageing, bearded and balding, but nevertheless demonstrating the prospect of life to come? Perhaps he would look up to me as I looked up to my own father, and perhaps he would welcome the transition. But perhaps not. There is clearly a risk. I fear that a child brought into a family as a clone could severely compromise family life – and that the child himself or herself would suffer the most. It surely is not proper to bring a child into the world in the expectation of such confusion and possible unhappiness.

What of the second commonly mooted motive for human cloning – to replicate very special individuals? Mozart, Einstein and Gandhi have commonly been proposed as candidates. The decision in this case would presumably be made not by an individual, wanting to regain a loved one, but by society; although we would have to envisage that the child must be brought up in a loving family.

Then again, for all the reasons already outlined, a genomic clone may not share the talents, and certainly not the inclinations, of his or her clone parent. A cloned Einstein would almost certainly be highly gifted, but he would not necessarily be a great physicist. To what extent, too, is it possible or desirable to simulate the upbringing of the original, which obviously had such an influence? Newton, one of the supreme talents of all time, had an extraordinary and unhappy childhood. He was abandoned by his mother at the age of three, although she continued to live in the next village with a new husband. Did the loneliness of this gratuitous separation contribute to the intensity and obsessiveness that later made him so remarkable? It is certainly possible. Should a Newton clone be subjected to similar cruelty? Surely not. Would a Newton clone brought up with siblings and taken out on picnics like any other child become so dedicated? Perhaps. Perhaps not.

There is of course another side to this particular coin: the hypothetical cloning of crazed dictators, effected at their own command or through the offices of their followers. In the days and weeks that followed Dolly's birth, many a commentator raised this possibility as one of several 'worst-case scenarios'; hence the battalion of Hitlers on the cover of *Der Spiegel*. But a cloned dictator would not replicate the originals any more than a cloned genius would. It has often been suggested that Hitler might have become a perfectly innocuous landscape painter if only the Vienna Academy of Fine Arts had accepted him as a student before World War I, and we might reasonably hope that the Hitler clone would be luckier in his choice of university.

A genetic, nuclear clone of Hitler would not necessarily strive to create a Fourth Reich. In reality, indeed, this would be most unlikely. If he inherited his clone father's oratorial powers he might as soon be a schoolteacher – or a priest, which was one of Hitler's own boyhood ambitions. If he was as fond of dogs, he might become a vet. Of course, the clone's genetic inheritance would set limits on his achievements. Richer, post-war nutrition would surely ensure that a Hitler clone was taller than the original but still he would never shine at bastketball, or trouble the Olympic scorers in the high jump. Unless he grew a pokey moustache and smeared his hair across his forehead, few would spot the resemblance to his infamous father. The Fuehrer's cloned offspring would surely disappoint their clone pater no end.

The cartoonists' vision of an instant battalion of Hitlers is a further nonsense. Clones like Dolly may be produced from adult cells but they begin their lives as one-cell embryos, and then develop at the same rate as others of their kind. Adolf Hitler was 44 years old when he became dictator of Germany and 50 at the outbreak of World War II. It would take just as long to produce the *doppelgänger* as it did to shape the original; and by that time the political moment that brought the first Hitler to power would be well and truly past, as indeed it is already.

It is always worth raising 'worst-case scenarios' when contemplating the future. We should examine every idea we can think of. But it must be a mistake to take worst-case scenarios at their face value, or to base general policy upon them. As the lawyers say, 'hard cases make bad law'. We have made clear by now that we are not apologists for human cloning. We do believe that all possible caveats – all that anyone can think of – should be laid out for discussion. It is the case, though, that some at least of the bogeys, when confronted head on, prove to be insubstantial.

Finally, however, there is one worst-case possibility that might have some substance. In the film *Bladerunner*, the cloned replicants were shown as becoming steadily more flawed as the generations pass, just as successive photocopies lose quality. Taken literally this is a nonsense – clones are not photocopies – yet the general notion does give cause for concern. Sexual reproduction ensures that new individuals are made from the fusion of gametes – egg and sperm. Gametes are produced by meiosis – 'reduction division' – as described in chapter 5 and elsewhere. In meiosis, it seems likely that any damaging mutations that arose during the life of the individual are shuffled off; indeed this might be the chief selective pressure that drove meiosis – and hence sex – to evolve in the first place. But clones are not produced by fusion of gametes. There is no meiosis in the prefatory phase to purge damaging mutations. Any mutations that are present in the particular cell nucleus from which the clone is made will be present in all the cells of the new individual. Cloning so far has proved difficult: most of the reconstructed embryos die. Mutation within those cells *could* be a cause. This has yet to be investigated. The survivors could just be the lucky ones – but they might still carry mutations of a sub-lethal nature. But if the cloned individual was cloned again, and then again and again, the results could well be increasingly disorderly.

In practice, this scenario is surely unlikely. We reproduce our own cloned animals by normal sexual means, and we may presume that cloned human beings (if ever there are any) would also reproduce by sex. It is hard to envisage circumstances in which a cloned individual would then be cloned again. If this did happen, however, then we could envisage an accumulation of flaws, much the same as in *Bladerunner*.

But we are now well in the realms of fancy and should return to the final, most important, more down-to-earth reason why human cloning has been seriously mooted. It could in principle be employed as an aid to infertility. Here again there are two clear-cut possibilities. The first is to help heterosexual couples in which both partners are infertile and who cannot be helped by any other means; and the second is to extend the traditional concept of 'family', and of 'parent'. Again, we do not advocate human cloning in this context, or any other. But the arguments should be explored.

Cloning to help infertility

The techniques outlined at the start of this chapter are intended to help couples who are unable to reproduce sexually because of some specific shortfall: lack of sperm, a defective womb, or whatever. In some cases, however, reproduction by sex is out of the question: when one person alone seeks to reproduce, without a sexual partner; or when two partners either fail to produce gametes at all, or produce incompatible gametes – as is the case with homosexual partners. In mammals, new embryos cannot be created by placing two sperm nuclei or two egg nuclei in a single egg because of the phenomenon of genomic imprinting: nuclei of both sexes are needed. But male homosexual couples might conceivably be cloned with the aid of egg donors and surrogate mothers, while female couples could be far more independent; indeed cloning could help to realise the dream of some extreme feminists – of reproduction without males. One member of a lesbian couple might provide a nucleus from a body cell, and another provide the cytoplasm; and next time around they could reverse the procedure. Of course, by such means, a lesbian couple could produce only daughters. A woman could clone herself precisely if one of her own nuclei were introduced into one of her own enucleated oocytes. Many combinations can be envisaged.

It is certainly not up to us to comment on where the limits of parenthood ought to lie. Clearly, however, society has become more liberal. In the decade up to the mid 1990s fifteen US states allowed adoption by gay couples, and what any one state allows the other states are constitutionally obliged to recognise. Lee Silver estimates that up to six million children in the United States may now be living in families of 'same-sex' couples, either male or female. Note, though, that mere liberality is not enough if homosexual couples are to produce their own genetic offspring. Various third parties – clinicians, technicians and sometimes donor mothers (to provide egg cytoplasm) and surrogate mothers – must lend their active assistance. In this context John Robertson, an American lawyer and moralist, has coined the expression 'collaborative reproduction'; although we could say that this is a matter of degree since human birth in general is 'collaborative' as it is usual and desirable at least to involve a midwife.

Even so, is cloning justified? We believe it is not. Keith in particular stresses that he can envisage no circumstance that could not be aided simply by adapting present techniques. Male homosexuals *could*, for example, borrow or buy the services of egg donors and surrogate mothers, to produce a genetic facsimile of at least one of them. Lesbian couples already have recourse to anonymous sperm donors. Heterosexual couples who are doubly infertile could buy or borrow the services of egg donors and sperm donors to produce a child by IVF and, if necessary, raise the child within the womb of a surrogate mother. There is no circumstance in which a couple would absolutely *need* to produce a child by non-sexual means. Of course, a child produced with the help of an egg-donor or sperm-donor is genetically related to only one of the prospective parents (unless the donor is related to that parent). But, Keith asks, why is genetic relationship considered to be such a big deal? What truly matters is that a child, however produced, should be brought up within a loving home. If this is taken to be the criterion of acceptability, then there can be no circumstances in which cloning is the only option.

Keith also suggests that doctors in the future might see cloning in the manner that produced Dolly as a crude, late-20th-century intermediate technology. As we have seen, it is already becoming possible to direct the differentiation of cells in culture – to induce stem cells to develop into particular tissues. Within a few decades such techniques will surely be standard. If this is the case, then it surely will be possible one day to induce cultured cells to develop into eggs or sperm. When this technology comes on board, it will be possible to produce embryos by IVF, from gametes made from body cells. These embryos could then be taken to term by surrogate mothers. This technology will not be cloning, although it will partake of the methods now being developed through cloning research.

Whatever way you turn the arguments, then, we can see no unequivocal reason to adopt human cloning, but we can envisage many shortcomings. It is the case, though, that cloning *qua* cloning is not the only issue. As we have stressed throughout this book, we have developed cloning technology primarily as the means by which to realise the potential of genetic engineering. So we should look at this, too.

People engineered

Cloning and genetic engineering are conceptually linked because they are technically linked. The transgenesis of zygotes is a hit-and-miss affair, offering only limited possibilities; but when cells are cultured by the million, and laid out in a dish for months at a stretch, genetic engineers can work their full repertoire. Then, the cells that are satisfactorily transformed can be made into new individuals by the same method that enabled us to create Polly. Thus, genetic engineering is a quite different concept; but cloning is implicated within it. Genetic engineering also seems to raise much bigger issues, for cloning merely duplicates what exists already – though only up to a point – while genetic engineering can in principle change the nature of living creatures; which means it might change the nature of humanity, the meaning of *Homo sapiens*. All this is feasible – but how could it come about? By what route, and with what motivation? When?

Genetic engineering might be applied to human beings in two quite different contexts. In the first, damaged tissues might be removed, manipulated, and then returned. Alternatively – and this is the approach that I am already involved in with Roslin Bio–Med – each person might be provided with his or her own personal store of cultured fetal cells that might, when required, be cultured and differentiated *in vitro* to replace or supplement damaged tissue of whatever kind is required. Such approaches are already being worked upon for a diversity of disorders, from cystic fibrosis to Parkinson's disease. This is high-tech, 21st century medical biology, but it does not seem to raise outstanding ethical issues. The physical changes made by such means are not passed on to the next generation. The only serious caveat seems to be that in some circumstances transformed cells intended for one person might escape into another – which at least is theoretically possible if such cells were delivered to the lung by aerosol, for example. But this is a technical issue, and surely resoluble.

The second possibility does raise ethical issues: adding, subtracting, or altering genes in the germ-line, so that future generations are affected. If an animal (including a human animal) is genetically transformed as a zygote, or by the method that made Polly, then the transformations should finish up in the eggs or sperm, and be passed into the dynasty. Whatever the nature of the gene – even if it is the most innocuous conceivable, even if it has no discernible effect at all – this is a serious thing to do. Tom Payne suggested at the end of the 18th century that no one generation had a right to impose its will on the next – though in practice this is unavoidable and, at least in some instances, future generations can reverse the decisions of their predecessors. But to add genes to future generations – that changes the nature of future people. It is hard to imagine a greater imposition. We might indeed simply leave the matter there and declare that germ-line transformation of human beings is in principle forever beyond the pale.

But every possibility should be thought through – and in fact, some serious physicians are already envisaging cases where, they suggest, germ-line transformation would be justified. So we should look further.

To begin with we can draw the traditional distinction that runs through all medicine: on the one hand, the correction of obvious disorder; and on the other, the enhancement of what already works reasonably well. In this way physicians distinguish between medicines and tonics. The distinction is not always easy in practice, but in principle it seems clear enough. Also, as we discussed in the context of Tracy, we should distinguish between characters that can be modified cleanly and discretely by adding, knocking out, or altering a single gene, and those that are polygenic, where each gene is merely a player in a larger consortium; and bear in mind, too, that most genes are pleiotropic, often having many effects besides the one that is intended.

An obvious target is to correct the many different 'single-gene disorders' – at least 3000 are known in human beings – that are caused by single mutated alleles. Robert (Lord) Winston has suggested Duchenne muscular dystrophy as a prime target. It is a sad and crippling disease: those who inherit the causative gene in double dose are increasingly disabled and die young.

There is of course an alternative approach. Most of the deleterious alleles that cause single-gene disorders in human beings are recessive. They have no obvious effect unless inherited in double dose, from both parents. Individuals who carry only one copy of the 'bad gene' are carriers, and have a fifty–fifty chance of passing the gene on to any one of their offspring, but they are not themselves affected. The simplest mendelian rule shows that if two carriers of Duchenne muscular dystrophy (or cystic fibrosis or sickle cell or whatever) have children, then two out of the four are liable to be carriers themselves (having inherited the bad gene from one parent but not the other); one out of four will inherit a double dose of the bad gene, and so will manifest the disorder; but one out of every four will inherit a normal allele from each parent, and so be totally free of the disorder altogether – neither affected, nor even a carrier. Of course these figures are averages: lucky parents who were both carriers might have half a dozen children totally free of the disease or they might have one affected child after the other.

Be that as it may: more and more of the mutant alleles that cause such disorders can now be diagnosed within very young embryos simply by taking a single cell and examining its DNA, by techniques that are becoming ever more simple and sure-fire. So in families where there is a history of disorder, and therefore some reason to look for trouble, it should soon be relatively simple to produce a batch of embryos by superovulation and IVF and to pick out the ones – one in four of them – that is free of the bad gene altogether. Then this one can be implanted into the mother.

In practice, however, I see nothing wrong ethically with the idea of correcting single-gene defects. But I am very concerned about any other kind of intervention, for anything else would simply be an experiment. How could we possibly foretell the outcome? Thus if we practised germ-line intervention *except* to correct very special defects that are well understood, we would not only be imposing our will on future generations, we would also be taking a chance – and potentially a very serious one – with their welfare. The chanciness alone seems to me to put such intervention beyond the pale. This brings us to an issue that must run through all debate and prophecy, whether of cloning *qua* cloning, or cloning as a means to genetic engineering. The universal issue is that of risk.

Risk

Risk is a tricky concept. It clearly has ethical implications: likelihood of success or of hideous failure must influence considerations of right and wrong. It is also one of those ethical regions to which science must contribute, for risk should be quantified and science alone can provide the necessary data and maths. But it is hard to get a feel for risk. We may estimate by the best possible means that such-and-such a procedure has a one in a hundred chance of causing such-and-such a disaster and feel that such a risk is worthwhile; but then we may find that the first three attempts are disasters. If the estimate was right then the chance of three failures in a row is one in a million; but one-in-a-million odds sometimes come up. Risk means that the outcome of any one endeavour is uncertain; and, of course, except in well-defined circumstances where we can enjoy the thrill, we greatly prefer certainty.

But then there are risks that can be quantified – because we have a great many examples from the past by which to make judgements, and/or some theory to tell us what ought to happen; and risks that cannot be quantified – because we simply don't have enough data to make a judgement; and risks that are yet unknown (a risk of risk). Then we must balance small risks of terrible disasters (in this context, the risk of a child being seriously deformed, for example) against large risks of mere inconveniences. Always we must exercise judgement. There are no absolute criteria of acceptability.

We do know, though, that risk cannot be avoided altogether even at the best of times. All action is potentially risky yet we have to act in order to survive: Miss Havisham's life of irreducible inactivity in *Great Expectations* probably did not extend her life (house-mites must have played havoc if nothing else) and certainly was not life enhancing. Babies conceived in the best circumstances sometimes die or are deformed; and this unpleasant fact does at least provide us with a yardstick. The risk of death or deformity in a cloned baby born to a surrogate mother can be measured against the risk of such disasters in more natural circumstances. But, in addition, we must balance the risk of action against the known disadvantage of inaction. Desperate couples faced with infertility have already shown that they are prepared to run considerable risk. People who know they carry harmful genes often risk having children. Perception of acceptability varies with circumstance as indeed it must.

How can such discussion throw light on the risk of cloning a human being by the method that produced Dolly? Work on sheep and mice already suggests that the chances are low of producing an embryo that is able to go to term from a given cell: Dolly was the only success out of 277 reconstructed embryos. Such statistics, however, do not tell us all we need to know. Perhaps we were unlucky: perhaps if we had made ten more embryos, we would have had ten more Dollys. Perhaps we were lucky: perhaps another 10 000 embryos would have yielded no more Dollys at all. Much more experience with more animals will give a clearer idea of the odds and also of course improve the rate of success – for practice always does reduce imperfection although, unfortunately, the reasons for the improvement are not always clear. All the results so far at Roslin suggest that, by present techniques, fetuses produced by nuclear transfer are ten times more likely to die *in utero* than fetuses produced by normal sexual means, while cloned offspring are three times more likely to die soon after birth. Deformities also occur. These setbacks are very distressing for people who work with animals – and surely mean that any immediate extension into human medicine is unthinkable.

But then, with Dolly, there is a list of unresolved issues that cannot at present be quantified – and if they could, what should we make of them? Dolly's genes came originally from a six-year-old animal and although Dolly after three years looks like an eminently normal three-year-old, we have yet to see how her life will pan out. The telomeres on her chromosomes are reduced in length. The shortening of the telomeres does not *cause* ageing, it merely reflects ageing; but when the telomeres have disappeared altogether, the cell does die. There is no reason to suppose that Dolly's telomeres have shortened enough, or will do so, to affect her lifespan but we cannot foresee how a longer-lived animal like a human being would be affected. Neither does Dolly seem to have suffered side-effects from genomic imprinting – but her embryo siblings, who did not make it to term, may have done. Certainly the potential issue of genomic imprinting is not resolved and it is hard to see how it can be until the phenomenon itself is better understood. Then there is the matter of large fetus syndrome. This is clearly a hazard in cattle and is certainly a potential hazard in sheep even though, because of the cross-breeding that has gone on, it is impossible at present to assess the extent of the problem in sheep. Yet these are only the potential hazards that we know about – or at least have some intimation of. Others will surely crop up that cannot at present be anticipated.

How can all the potential hazards be identified and quantified so that we know in advance what the risks would be, if anyone did ever attempt to clone a human? They can't, is the answer. Dolly and her daughter Bonnie must live their allotted spans and perhaps their early promise of total normality will be confirmed. The Hawaiian cloned mice have already bred through several generations and will generate far more data, far more quickly. Other species surely will be cloned before humans are – they certainly should be – such as dogs and, of course, monkeys, which are primates like us. But still, we cannot know until we try that the lessons learned from all those other species are strictly applicable to us. As Keith has abundantly observed, there are huge similarities in the biology of different creatures that stretch from humans to yeasts and well beyond. But there are many differences of detail, too, as he also stresses; and details can make all the difference. At present there is no obvious reason to suppose that human cloning will raise problems that could not be foreseen in sheep or mice or monkeys. Besides, human reproductive biology has been well studied; and already remarkable manipulations are possible. So human cloning might prove to be reasonably straightforward, once biologists have gained experience in a range of other species. But nobody will know until they try.

Whenever we speak of risk, too, we must ask 'Risk to whom?', and 'Risk of what?' Our first thoughts turn to the physical wellbeing of the baby – will it live, will it be normal? But others are physically involved too, notably the surrogate mother – and, so many would insist, the reconstructed embryos who are not implanted, or are implanted but fail; and many people will be involved psychologically, including the baby who becomes the child, its social parents who bring it up, its surrogate mother, and so on.

So what is the risk of cloning people? At present the question is unanswerable even if we confine discussion to the physical wellbeing of the newly cloned individual. There are too many unknowns. How long would it take to reduce the unknowns to the point where the transition to humans seems acceptable? There is a 'How long is a piece of string?' quality to this question. The transition will always be a risk. The procedure is bound to carry some hazard, which seems likely to be higher than that of normal conception and childbirth; and the data from animals, however exhaustive, cannot in the end tell us all we would like to know. If, however, we were to insist that a range of species should be examined, and that cloned animals and their offspring should be allowed to live their complete lives, and that the hazards we can anticipate already (like those that might ensue from genomic imprinting) should be understood more thoroughly, then surely we should be thinking in terms of decades.

But some people are impatient. Some are already offering cloning, or proposing to do so. Richard Seed is not the only one. I told a government committee that 'significant progress' towards human cloning could be made within three years – and this was reported as if I had said that it will, in fact, be technically possible. Actually it might be technically possible in that time – but only in the sense that if people who are prepared to take risks and are competent to attempt it, then if they are lucky, they might get away with it. But 'getting away with it' is not what is normally meant by offering a clinical service. In the end, then, the date of the first human cloning will be determined by the amount of risk that potential parents, and those who attend to them, are prepared to take. Perhaps clinicians should warn their patients not to take the risks – just as we now advise 16-year-olds that they ought to wear crash helmets, even when they would rather not.

But advice is not always taken and clinicians do not always behave like this – so is human cloning, in the end, inevitable? The market forces are certainly strong, but nothing is inevitable. At present human cloning is illegal in Britain and much of the United States, but mere illegality is not quite the issue. All that is required in the initial stages, to produce a clone like Dolly, is a flask of cultured body cells. This could be prepared in any competent laboratory. A technician merely needs to take a tissue biopsy, and from that create a culture. Once the culture is growing in a stable fashion it can be frozen and transported anywhere in the world. The only laws that seem to apply so far are those that restrict transport of biological materials; but if they do apply in this case, it surely would not be difficult to get around them. The cultured cells could then be sent to a country where cloning (or genetic manipulation) is not illegal. This could be anywhere: anywhere, preferably, where local women are willing to act as surrogate mothers. In any case, the person who wanted to be cloned could simply travel to the appropriate laboratory, like any tourist. In short, a practice that is illegal in one country can become a minor industry in another.

Would people go to such lengths? The answer seems obvious. Many people are not against human cloning and have little regard for the laws that now restrict it. In all walks of life, laws that are not respected are routinely flouted. Many physicians would be perfectly willing to offer such a service. Many doctors certainly feel that they should make ethical choices but others argue that it is not up to them to decide what is right or wrong; their duty is merely to meet their patients' requests. Many would argue that it is morally wrong to withhold their services from people they could help – especially if they see nothing wrong with the procedure in the first place. Would patients be willing to take the risks and pay the price? Undoubtedly. There is no stronger human instinct than to reproduce. People in America routinely pay $50 000 for a course of IVF, with no guarantee of success. In the present culture of the USA, the marketplace leads, and there surely is a huge potential market for cloning: eager, affluent buyers; willing and capable suppliers.

There is a broader point, perhaps best illustrated by reference to the United States. In the western world we generally subscribe to the notion of democracy, and also tend to support the idea of free enterprise. The philosophy of democracy differs enormously in different parts of the world. In Britain, for example, it tends to mean 'majority rule' – although Britons pay increasing attention to the demands of various minorities. But some at least of the founders of the modern United States saw majority rule in an unsavoury light. Much more important, they felt, was personal freedom: the right of an individual to do what he or she wanted. Personal liberty remains the guiding light of the USA. Thus it is that, in that country, people who themselves object to the idea of human cloning are quite likely to defend the rights of people who welcome it. It is always difficult to maintain unpopular laws in democracies. In societies where personal liberty is perceived effectively to be sacrosanct, it is surely almost impossible.

On the face of things, then, it may seem that human cloning (and perhaps, from this, germ-line engineering) is very likely to happen: perhaps in the first half of the 21st century, if not in the first decade. Yet there are further points, which suggest that in reality it is far from inevitable; or at least, that it will never have anything like the impact that, say, IVF has had. First, various societies in recent years have shown that they *can* resist new technologies of many kinds, whatever the market forces. Thus various European countries have now rejected nuclear power. The forces massed behind the nuclear industry are prodigious; but the Swedish people, among others, have said they simply do not want it. In Britain, popular feeling has effectively condemned the tower block, which as recently as the 1970s was seen as the only efficient form of mass housing, and had huge financial backing. Spectacularly, as the new millennium dawns, there is widespread rejection on both sides of the Atlantic of genetically modified – 'GM' – crops. In general, too, the world over, we may perceive a trend towards libertarianism. But countries do become more restrictive, as well as less. The USA and Europe now have laws to protect wildlife and to restrict the sales of many different items – including drugs – which 100 years ago would have been seen simply as an affront to freedom. Similarly, we need not assume that future societies will inevitably become more relaxed about human cloning than we are now; and, even if they did, their successors may later decide that enough is enough, and set themselves against it. There are precedents. There is no inevitable 'slippery slope'.

Then again, the extent to which people do *not* adopt novel technologies is striking. AI has been available for 200 years. It is far more common now than in the past yet it remains very much a minority pursuit. Some extreme feminists have envisaged 'families' of women, living like prides of lionesses, employing men (or AI) simply for their genes and raising the children in creches. Why should a woman tie herself to some fickle and self-indulgent couch potato when such freedom is so easily achieved? Why indeed? Well, social conservatism is obviously one answer. But another, perhaps, is that women prefer to live in a family with a single mate, however imperfect he may be. Human beings are not quite so flexible as it has become fashionable to argue. Some ways of life come easier to us than others. By the same token, human beings are sexual beings. Other routes do not come naturally to us. Of course, as moral philosophers remind us, what is 'natural' is not necessarily right, and what is 'unnatural' is not necessarily wrong. It is true, however, that what comes naturally to us also tends to come easily. When it comes to the point, many people surely will find themselves averse to cloning; and perhaps that aversion will prove to be deep seated; and perhaps, however great the market forces or the desire to reproduce, that aversion will prevail. This cannot be predicted. We will just have to wait and see. Clearly, though, there is reason to doubt whether human cloning will ever become fully respectable, let alone fashionable; and it surely will never be the norm. It may seem crude but it is surely accurate to suggest that reproduction by cloning goes against our nature.

Overall, as with all powerful technologies, there are contradictions. On the one hand, cloning has immense potential for good – especially in combination with the other great modern biotechnologies, genetic engineering and genomics. But such power can be abused and the most obvious possible abuse – at least, we believe it to be so – would be human cloning. The pressures for human cloning are powerful; but, although it seems likely that somebody, at some time, will attempt it, we need not assume that it will ever become a common or significant feature of human life. Society does not have to adopt technologies with which it feels uncomfortable; and people have already shown that they are perfectly able to resist what they do not like.

Finally, what does the future hold for us?

Moving On

Colin Tudge:

So where are Ian and Keith now, and how do they intend to spend the next few years and, indeed, the rest of their careers? What will happen to cloning – better known these days as nuclear transfer or NT technology – and what part will they play in future developments? What do they think, more broadly, of the grand events, the sea-change in biotechnology and hence in the world's ambitions, that they have helped to bring about?

Whatever they achieve, they will not achieve it together – at least in the immediate future. Theirs has been one of the most fruitful partnerships in modern biology, but it is a partnership no longer. They now work in different institutions, and insofar as they are working on the same problems, they are in effect rivals. In gentler times their separateness might not have mattered: they might have shared day-to-day thoughts in the name of science and academic freedom. These days, however, commerce prevails. Patents come first; discussion comes later.

At a personal level Keith says, with some emphasis, 'I don't want fame; but I do feel that my contribution to the science and management of the projects that led to Dolly were significant, and some recognition would be nice.' He has that recognition, however, for on 2 August 1999 he wrote to the vice chancellor of Nottingham University to accept a chair in the school of animal physiology, with £250 000 for equipment and the opportunity to raise several millions more in grants, and so to establish a fine new department of his own. The one-time technician with a low boredom threshold is now Professor Campbell. He also keeps a foothold at PPL: a working arrangement as nearly perfect (an outsider might be forgiven for suggesting) as might be envisaged. Nottingham is one of the world's great centres of reproductive biology (and, of course, is Ian's alma mater).

Keith's ambition, he says, is 'to go on enjoying science'; and specifically, 'I want to be the first to clone a pig!' Pigs are important in this context not as livestock, but because they are prime candidates to provide organs for transplantation into humans – that is, for 'xenotransplantation'. As for many companies, xenotransplants are the next great target for PPL. So why are pigs so difficult? – 'Well,' says Keith, 'if I knew that they wouldn't be!' In truth, though, reproductive technologies in general have proved difficult in pigs. For example, their embryos will not easily survive freezing. The high lipid – fat – content of the oocytes and young embryos is clearly a factor. Embryo transfer is complicated because sows in general need to carry at least five fetuses to sustain a pregnancy. These are technical details; there is no fundamental reason to suppose that the reproduction of pigs cannot, in the fullness of time, be controlled as readily as sheep. But the technicalities still need to be overcome.

But of course, the mere cloning of pigs is only the starting point. Although pig organs are similar in size and general function to those of humans, they are very different in the details of their cell surfaces and would be massively rejected if transplanted in their native state. Genetic transformation is needed: to add genes to produce surface proteins that will prevent the hyperacute response, and to remove the genes that produce the antigenic proteins that provoke the more prolonged immune responses. The technology that produced Polly shows the general route: grow the pig cells in culture, transform them genetically as they develop, and then create embryos from the nuclei that show the appropriate genetic changes. So far, however, it is not possible to clone pigs even in the way that Megan and Morag were cloned – even without any genetic transformation.

But the science and technology of genetic engineering also need to be advanced if the appropriate changes are to be made reliably. Gene targeting is needed – placing the required genes at the exact point in the chromosome where they will best be expressed and controlled. 'Gene targeting,' says Keith, 'is the Holy Grail of genetic engineering.' But it seems that this requirement, at least, is now being met. On 22 June 1999 PPL Therapeutics plc announced that they had achieved gene targeting in two lambs, Diana and Cupid – the latter, in his original state of quasi-divinity, known for the precision of his bowmanship; and, say PPL, there are more to follow. But neither the managing director, Ron James, nor the research director, Alan Colman, would say what genes they had transferred, or where they had been positioned in the receiving chromosomes, or how the trick had been brought about. They made the announcement not for the benefit of their fellow scientists, in the traditional spirit of academe, but for PPL's shareholders. In fact PPL may withhold formal publication in a learned journal until the appropriate patents are cleared. Things have moved on even since the days of Megan and Morag.

Ian, since Dolly, has become famous – 'though more in America than in Britain,' he says. He has delivered scores of lectures, to pensioners in Florida, campuses of Catholic students, and many more besides; and has helped committees of scientists, politicians and 'ethicists', including the Senate Committee in Washington where he met Edward Kennedy, and Britain's parliamentary committee in London. 'I expected the level of interest to decay,' he says, 'but it has persisted at a much higher level than I anticipated.' His work-load is horrendous and, as he commutes across time-zones, must largely be accomplished without sleep. The honours that have come his way include a Doctorate of Science (DSc) from Nottingham, and another from the NorthEastern University in Boston, Mass., a large private university that does much to educate students from poorer backgrounds. He continues to work at Roslin, and with Roslin's own biotech company Bio-Med – now merged with the California company, Geron. 'Things are much more dynamic these days,' says Ian. 'I like that.'

High among Geron's immediate priorities is what is now known as 'therapeutic cloning': culturing and reprogramming the patient's own cells to make healthy replacements for dysfunctional tissues – pertinent to patients with conditions as common and diverse as Alzheimer's disease or diabetes. This work relates most directly to the experience that helped to draw Ian into his present research: the effect of diabetes on his own father.

Both Ian and Keith, in all their capacities, continue to be interested in basic science; the mechanisms that underlie their technical achievements. Only when the mechanisms are fully understood is it possible to 'optimise' the technologies, to make them as powerful and precise as they can be. Again, though, we see the dialogue between science and technology – for the technology of nuclear transfer and genetic transformation themselves will lead on to the scientific insights on which better technologies can be built. Those who contemplate human cloning would do well to note that, at present, the successful technologies run far ahead of 'true' understanding.

Both Ian and Keith – like all biologists – would like fully to understand the greatest miracle of all: how a single cell with its coils of DNA divides and divides again to become a frog, or a sheep, or a human being; so both will continue to work on various aspects of development. The technology of nuclear transfer clearly provides a wonderful investigative tool. Both, too, seek to understand the phenomenon that lies at the heart of their success with Megan and Morag, and then with Dolly: the extraordinary fact, so long denied, that the genomes of differentiated cells can be reprogrammed to the extent that they can then instruct the development of an entire, perfect animal. Once we understand reprogramming then, surely, we will understand differentiation itself. Put that together with a detailed knowledge of development and we will at last have a solid answer to Kipling's just-so question, 'How does an animal become?'

There is a huge amount to do: scores of details, each of which may take years to work through – first in one species, then in several, and then in enough to be able to make broad generalisations about the range of possible mechanisms and the likelihood that any particular one of them will operate in any particular circumstances.

At Nottingham, for instance, Keith intends to continue research with colleagues at Nottingham and elsewhere, on the maturation of oocytes, both in cattle and in humans: why, for example, in the ovary, one particular follicle is selected for maturation and extrusion. At the moment, many facts are known: that such-and-such a thing happens at such-and-such a time. *Why* they happen in the way they do, in preference to some other possibility, is much less certain.

Biotechnology in general is among the most powerful of all the legacies that the 20th century has left to the 21st. The chief of the biotechnologies is genetic engineering, the transfer and alteration of genes; and the true significance of cloning is to enhance genetic engineering in animals – to make it a feasible procedure, routinely and reliably. PPL's precisely modified lambs, Cupid and Diana, have already taken genetic engineering into its next phase. Ian, who began the programme that led to Megan and Morag and beyond, and Keith, who provided the insights that in the end made it all work, have been and are key figures. They have no doubt whatever about the good that their work can bring; but they do of course have anxieties, too. Their most immediate and obvious cause of disquiet is human cloning, which has so comprehensively taken the attention of the world at large. They both, as we have seen throughout this book, despise the notion of it. Keith suggests that human cloning may already be possible technically – in the sense that somebody who is prepared to take enormous risks might just get away with it. But he stresses that the present obvious dangers – the high rate of late abortions and perinatal deaths, and the attendant deformities – will surely prevent anyone who is not actually deranged from making such an attempt in the immediate future. 'I suppose it will happen eventually, though,' he says. Whether human cloning catches on, of course, is another question; and by the time it truly becomes feasible it might already be obsolete as a reproductive technique (if, for example, it becomes possible to convert body cells into gametes). In the meantime, Ian and Keith fear that legislation designed to prevent human cloning may inhibit NT technology in general, including the 'therapeutic cloning' of body tissues.

There is another, broader aspect. Science and technology change our lives more directly and in the end more profoundly than any other social force. Governments do not so much direct technologies, as adjust to them; the political systems that come to prevail are the ones that adapt most adroitly to the most powerful techniques. To some extent new technologies inhibit our lives – television really has overridden many a local custom, for example – but to a much greater extent they expand the possibilities; enable us to do things that hitherto were unthinkable. So we must ask, how can we keep control of these technologies? It is not possible for everyone in a society to peruse every research proposal, so *who* in the end should have the final say? How in general can we ensure that on the one hand we do not kill the geese that lay the golden eggs – that we don't stop scientists from following their noses; but on the other hand that we are not lumbered with technologies that offend us, or lower our quality of life, or simply hand over life's controls to powerful companies? In the modern world, dominated by capitalist democracies, we always have to ask, what is the proper balance between free enterprise and government control? Science and technology are expensive and powerful and within them, this question is ever more pressing.

Cloning and the other biotechnologies are at the heart of all these grand discussions. Megan and Morag were born in 1995. Ian first became involved in the research that led to them in 1982, and henceforth followed the tortuous route we have described in this book; inevitably tortuous, because there was so much that could not be foreseen. Would such persistence and changes of direction have been possible except in a government or a university laboratory – one that was not driven by the need to placate shareholders? On the other hand Ian says that in recent years 'I am very aware of how much Keith and I have owed to venture capital'. In short, without entrepreneurs to seize the various initiatives, those grand ideas could have died on the vine. But now we see a major conceptual advance of fundamental importance – gene targeting, as manifest in Cupid and Diana – announced but unreported, the details withheld for fear of compromising patents. More broadly, in many contexts, people worldwide complain that the companies with the wherewithal to develop high technologies invest only in those that will bring obvious profit – techniques to alleviate the aches and pains of the rich and elderly, for example, rather than those that reduce infant mortality among the poor; but the companies in turn protest that they *have* to satisfy their shareholders, or go out of business.

So there is a lot for the 21st century to deal with. There is a huge amount of science to be done, and although the pace of advance is breathtaking, serious science takes decades or even centuries to unfold. Nearly half a century has passed since Francis Crick and James Watson first launched the science of molecular biology yet the technology that has derived from it, genetic engineering, is still in its infancy. There is a lot of ethical thinking to be done, too: present discourse leaves much to be desired. There are huge political issues that will never go away, for they seem to be part of the human condition: who should set the agenda; who should be in control; who should control the controllers, and how.

Ian and Keith both say, 'I want to carry on doing good science'; and each suggests that their social responsibility, primarily, is 'to explain what the science is about, as clearly as we are able'. Keith is disappointed with the press: 'They have gone for headlines, started at the sci-fi end of things, not focusing on the benefits.' On the other hand, there is now a children's book on how to clone a sheep – and, says Keith, 'It's all in there!'

So it is not inevitable that science and high-tech should get out of society's control. Science can be very difficult and some of it – like quantum mechanics – can be understood only by people with unusual ways of thinking and even they find it hard, which is why Niels Bohr and Werner Heisenberg had so many long conversations, over so many years. But biology is the science that affects us most profoundly and biology, in the main, is not counterintuitive. It is merely complicated, like chess or the plot of *As You Like It*: anyone can understand it if they put their minds to it. But unless people do understand science then democracy is mocked, because science and the technologies that flow from it are the most significant agents of social change. Biotech is among the most significant – perhaps the most significant – of all technologies, and cloning sits at the heart of it. The amiable Dolly already seems a little old-fashioned in this age of Cupid and Diana, but she is a symbol nevertheless of times to come. We hope this book has aided the necessary understanding. Science changes the way we all live, but scientists are citizens too.

The Letter to Nature Announcing Dolly's Birth

Viable offspring derived from fetal and adult mammalian cells

I. Wilmut, A. E. Schnieke*, J. McWhir, A. J. Kind* & K. H. S. Campbell

Roslin Institute (Edinburgh), Roslin, Midlothian EH25 9PS, UK
*PPL Therapeutics, Roslin, Midlothian EH25 9PP, UK

Fertilisation of mammalian eggs is followed by successive cell divisions and progressive differentiation, first into the early embryo and subsequently into all of the cell types that make up the adult animal. Transfer of a single nucleus at a specific stage of development, to an enucleated unfertilised egg, provided an opportunity to investigate whether cellular differentiation to that stage involved irreversible genetic modification. The first offspring to develop from a differentiated cell were born after nuclear transfer from an embryo-derived cell line that had been induced to become quiescent. Using the same procedure, we now report the birth of live lambs from three new cell populations established from adult mammary gland, fetus and embryo. The fact that a lamb was derived from an adult cell confirms that differentiation of that cell did not involve the irreversible modification of genetic material required for development to term. The birth of lambs from differentiated fetal and adult cells also reinforces previous speculation that by inducing donor cells to become quiescent it will be possible to obtain normal development from a wide variety of differentiated cells.

It has long been known that in amphibians, nuclei transferred from adult keratinocytes established in culture support development to the juvenile, tadpole stage. Although this involves differentiation into complex tissues and organs, no development to the adult stage was reported, leaving open the question of whether a differentiated adult nucleus can be fully reprogrammed. Previously we reported the birth of live lambs after nuclear transfer from cultured embryonic cells that had been induced into quiescence. We suggested that inducing the donor cell to exit the growth phase causes changes in chromatin structure that facilitate reprogramming of gene expression and that development would be normal if nuclei are used from a variety of differentiated donor cells in similar regimes. Here we investigate whether normal development to term is possible when donor cells derived from fetal or adult tissue are induced to exit the growth cycle and enter the Go phase of the cell cycle before nuclear transfer.

Three new populations of cells were derived from (1) a day-9 embryo, (2) a day-26 fetus and (3) mammary gland of a 6-year-old ewe in the last trimester of pregnancy. Morphology of the embryo-derived cells is unlike both mouse embryonic stem (ES) cells and the embryo-derived cells used in our previous study. Nuclear transfer was carried out according to one of our established protocols and reconstructed embryos transferred into recipient ewes. Ultrasound scanning detected 21 single fetuses on day 50–60 after oestrus (Table 1). On subsequent scanning at ~14-day intervals, fewer fetuses were observed, suggesting either mis-diagnosis or fetal loss. In total, 62% of fetuses were lost, a significantly greater proportion than the estimate of 6% after natural mating. Increased prenatal loss has been reported after embryo manipulation or culture of unreconstructed embryos. At about day 110 of pregnancy, four fetuses were dead, all from

Table 1 Development of embryos reconstructed with three different cell types

Cell type	No. of fused couplets (%)[*]	No. recovered from oviduct (%)	No. cultured	No. of morula/ blastocysts stage embryos (%)	No. of morula/ blastocysts stage embryos transferred[†]	No. of pregnancies/ no. of recipients (%)	No. of live lambs (%)[‡]
Mammary epithelium	277 (63.8)[a]	247 (89.2)	–	29 (11.7)[a]	29	1/13 (7.7)	1 (3.4%)
Fetal fibroblast	172 (84.7)[b]	124 (86.7)	–	34 (27.4)[b]	34	4/10 (40.0)	2 (5.9%)
			24	13 (54.2)[b]	6	1/6 (16.6)	1 (16.6%)[§]
Embryo-derived	385 (82.8)[b]	231 (85.3)	–	90 (39.0)[b]	72	14/27 (51.8)	4 (5.6%)
			92	36 (39.0)[b]	15	1/5 (20.0)	0

[*] As assessed 1h after fusion by examination on a dissecting microscope. Superscripts a or b within a column indicate a significant difference between donor cell types in the efficiency of fusion ($P < 0.001$) or the proportion of embryos that developed to morula or blastocyst ($P < 0.001$).
[†] It was not practicable to transfer all morulae/blastocysts.
[‡] As a proportion of morulae or blastocysts transferred. Not all recipients were perfectly synchronised.
[§] This lamb died within a few minutes of birth.

embryo-derived cells, and post-mortem analysis was possible after killing the ewes. Two fetuses had abnormal liver development, but no other abnormalities were detected and there was no evidence of infection.

Eight ewes gave birth to live lambs (Table 1). All three cell populations were represented. One weak lamb, derived from the fetal fibroblasts, weighed 3.1 kg and died within a few minutes of birth, although post-mortem analysis failed to find any abnormality or infection. At 12.5%, perinatal loss was not dissimilar to that occurring in a large study of commercial sheep, when 8% of lambs died within 24h of birth. In all cases the lambs displayed the morphological characteristics of the breed used to derive the nucleus donors and not that of the oocyte donor (Table 2). This alone indicates that the lambs could not have been born after inadvertent mating of either the oocyte donor or recipient ewes. In addition, DNA microsatellite analysis of the cell populations and the lambs at four polymorphic loci confirmed that each lamb was derived from the cell population used as nuclear donor. Duration of gestation is determined by fetal genotype, and in all cases gestation was longer than the breed mean (Table 2). By contrast, birthweight is influenced by both maternal and fetal genotype. The birthweight of all lambs was within the range for single lambs born to Blackface ewes on our farm (up to 6.6 kg) and in most cases was within the range for the breed of the nuclear donor. There are no strict control observations for birthweight after embryo transfer between breeds, but the range in weight of lambs born to their own breed on our farms is 1.2–5.0 kg, 2–4.9 kg and 3–9 kg for the Finn-Dorset, Welsh Mountain and Poll-Dorset genotypes, respectively. The attainment of sexual maturity in the lambs is being monitored.

Table 2 Delivery of lambs developing from embryos derived by nuclear transfer from three different donor cells types, showing gestation length and birthweight

Cell type	Breed of lamb	Lamb identity	Duration of pregnancy (days)*	Birth-weight (kg)
Mammary epithelium	Finn-Dorset	6LL3	148	6.6
Fetal fibroblast	Black Welsh	6LL7	152	5.6
	Black Welsh	6LL8	149	2.8
	Black Welsh	6LL9†	156	3.1
Embryo-derived	Poll-Dorset	6LL1	149	6.5
	Poll-Dorset	6LL2‡	152	6.2
	Poll-Dorset	6LL5	148	4.2
	Poll-Dorset	6LL6‡	152	5.3

*Breed averages are 143, 147 and 145 days, respectively for the three genotypes Finn-Dorset, Black Welsh Mountain and Poll-Dorset.
† This lamb died within a few minutes of birth.
‡ These lambs were delivered by caesarian section. Overall the nature of the assistance provided by the veterinary surgeon was similar to that expected in a commercial flock.

Development of embryos produced by nuclear transfer depends upon the maintenance of normal ploidy and creating the conditions for developmental regulation of gene expression. These responses are both influenced by the cell-cycle stage of donor and recipient cells and the interaction between them. A comparison of development of mouse and cattle embryos produced by nuclear transfer to oocytes or enucleated zygotes suggests that a greater proportion develop if the recipient is an oocyte. This may be because factors that bring about reprogramming of gene expression in a transferred nucleus are required for early development and are taken up by the pronuclei during development of the zygote.

If the recipient cytoplasm is prepared by enucleation of an oocyte at metaphase II, it is only possible to avoid chromosomal damage and maintain normal ploidy by transfer of diploid nuclei, but further experiments are required to define the optimum cell-cycle stage. Our studies with cultured cells suggest that there is an advantage if cells are quiescent. In earlier studies, donor cells were embryonic blastomeres that had not been induced into quiescence. Comparisons of the phases of the growth cycle showed that development was greater if donor cells were in mitosis or in the G1 phase of the cycle, rather than in S or G2 phases. Increased development using donor cells in G0, G1 or mitosis may reflect greater access for reprogramming factors present in the oocyte cytoplasm, but a direct comparison of these phases in the same cell population is required for a clearer understanding of the underlying mechanisms.

Together these results indicate that nuclei from a wide range of cell types should prove to be totipotent after enhancing opportunities for reprogramming by using appropriate combinations of these cell-cycle stages. In turn, the dissemination of the genetic improvement obtained within elite selection herds will be enhanced by limited replication of animals with proven performance by nuclear transfer from cells derived from adult animals. In addition, gene targeting in livestock should now be feasible by nuclear transfer from modified cell populations and will offer new opportunities in biotechnology. The techniques described also offer an opportunity to study the possible persistence and impact of epigenetic changes, such as imprinting and telomere shortening, which are known to occur in somatic cells during development and senescence, respectively.

The lamb born after nuclear transfer from a mammary gland cell is, to our knowledge, the first mammal to develop from a cell derived from an adult tissue. The phenotype of the donor cell is unknown. The primary culture contains mainly mammary epithelial (over 90%) as well as other differentiated cell types, including myoepithelial cells and fibroblasts. We cannot exclude the possibility that there is a small proportion of relatively undifferentiated stem cells able to support regeneration of the

mammary gland during pregnancy. Birth of the lamb shows that during the development of that mammary cell there was no irreversible modification of genetic information required for development to term. This is consistent with the generally accepted view that mammalian differentiation is almost all achieved by systematic, sequential changes in gene expression brought about by interactions between the nucleus and the changing cytoplasmic environment.

Methods

Embryo-derived cells were obtained from embryonic disc of a day-9 embryo from a Poll-Dorset ewe cultured as described, with the following modifications. Stem-cell medium was supplemented with bovine DIA/LIF. After 8 days, the explanted disc was disaggregated by enzymatic digestion and cells replated onto fresh feeders. After a further 7 days, a single colony of large flattened cells was isolated and grown further in the absence of feeder cells. At passage 8, the modal chromosome number was 54. These cells were used as nuclear donors at passages 7–9. Fetal-derived cells were obtained from an eviscerated Black Welsh Mountain fetus recovered at autopsy on day 26 of pregnancy. The head was removed before tissues were cut into small pieces and the cells dispersed by exposure to trypsin. Culture was in BHK 21 (Glasgow MEM; Gibco Life Sciences) supplemented with L-glutamine (2 mM), sodium pyruvate (1 mM) and 10% fetal calf serum. At 90% confluency, the cells were passaged with a 1:2 division. At passage 4, these fibroblast-like cells had modal chromosome number of 54. Fetal cells were used as nuclear donors at passages 4–6. Cells from mammary gland were obtained from a 6-year-old Finn-Dorset ewe in the last trimester of pregnancy. At passages 3 and 6, the modal chromosome number was 54 and these cells were used as nuclear donors at passage numbers 3–6.

Nuclear transfer was done according to a previous protocol. Oocytes were recovered from Scottish Blackface ewes between 28 and 33 h after injection of gonadotrophin-releasing hormone (GnRH), and enucleated as soon as possible. They were recovered in calcium- and magnesium-free PBS containing 1% FCS and transferred to calcium-free M2 medium containing 10% FCS at 37°C. Quiescent, diploid donor cells were produced by reducing the concentration of serum in the medium from 10 to 0.5% for 5 days, causing the cells to exit the growth cycle and arrest in G0. Confirmation that cells had left the cycle was obtained by staining with antiPCNA/cyclin antibody (Immuno Concepts), revealed by a second antibody conjugated with rhodamine (Dakopatts).

Fusion of the donor cell to the enucleated oocyte and activation of the oocyte were induced by the same electrical pulses, between 34 and 36 h after GnRH injection to donor ewes. The majority of reconstructed embryos were cultured in ligated oviducts of sheep as before, but some embryos produced by transfer from embryo-derived cells or fetal fibroblasts were cultured in a chemically defined medium. Most embryos that developed to morula or blastocyst after 6 days of culture were transferred to recipients and allowed to develop to term (Table 1). One, two or three embryos were transferred to each ewe depending upon the availability of embryos. The effect of cell type upon fusion and development to morula or blastocyst was analysed using the marginal model of Breslow and Clayton. No comparison was possible of development to term as it was not practicable to transfer all embryos developing to a suitable stage for transfer. When too many embryos were available, those having better morphology were selected.

Ultrasound scan was used for pregnancy diagnosis at around day 60 after oestrus and to monitor fetal development thereafter at 2-week intervals. Pregnant recipient ewes were monitored for nutritional status, body condition and signs of EAE, Q fever, border disease, louping ill and toxoplasmosis. As lambing approached, they were under constant observation and a veterinary surgeon called at the onset of parturition. Microsatellite analysis was carried out on DNA from the lambs and recipient ewes using four polymorphic ovine markers.

Received 25 November 1996; accepted 10 January 1997.

GLOSSARY

ACROSOME. The tip of the spermatozoon, containing enzymes that help the sperm to bore through the *zona pellucida* during fertilisation.

ACTIVATION. Various meanings, including the process by which contact with the sperm prompts the oocyte to complete the second phase of meiosis, and then to embark upon cell division.

ADENINE. One of the four nucleotides, or bases, that form the variable part of the nucleic acids, DNA and RNA.

ALLELE. A variant, or version, of a gene. (A polymorphic gene is one that is present in a population in more than one *allele*.)

ALPHA-1-ANTITRYPSIN (AAT). An enzyme that interacts with the enzyme *elastase* to maintain the correct elasticity of the lungs. In cystic fibrosis and emphysema the action of AAT is interrupted.

AMINO ACID. One of the 'building blocks' of a protein. A protein macromolecule consists of chains of amino acid molecules.

ANAPHASE. The phase of mitosis or meiosis in which the dividing chromosomes are driven apart.

ANEUPLOID. A cell (or organism) containing an anomalous number of chromosomes.

ANISOGAMOUS. Referring to the state in which the males and females produce gametes of different sizes, as in sperm and eggs.

ANTIBIOTIC. Literally, 'anti-life'. A chemical agent commonly (though not necessarily) produced by a fungus that destroys or inhibits other life forms, typically (though not necessarily) bacteria.

ARTIFICIAL INSEMINATION (AI). Introduction of sperm into the female tract for the purposes of impregnation, by artificial means.

ASEXUAL. Non-sexual.

BASE. An alternative term for 'nucleotide'.

BINARY FISSION. Division of a cell into two. The mode of reproduction favoured by many protists, including *Amoeba*.

BIOTECHNOLOGY. High technology of a biological nature.

BIVALENT. The groups of four chromatids that appear during meiosis, as homologous chromosomes come together prior to crossing over.

BLASTOCYST. An embryo at the stage when it is a hollow ball of cells; it is more advanced than a morula.

BLASTOMERE. A cell from an early embryo, up to the blastocyst stage.

BLASTOSPHERE, BLASTULA. The embryonic stage after the blastocyst, when the ball of cells becomes hollow.

CDC2. A protein that combines with cyclin to produce the protein kinase enzyme that is MPF.

CELL CYCLE. The complete life cycle of the cell, broadly divided into cell division (mitosis or meiosis) and interphase.

CENTROMERE. The structure in the chromosome to which the spindle attaches during mitosis or meiosis.

CHIMERA. A general term for any biological entity made up of components from more than one individual. In this context, primarily applied to embryos or offspring made up of cells from different embryos.

CHROMATIN. The total material of the chromosome including

the DNA and many different proteins.

CHROMOSOME. The physical structure within the cell nucleus of a eukaryote, which carries the genetic material in the form of DNA, wound around a protein core of histones.

CLEAVAGE. The final division of a cell; that is, the division of the cytoplasm which follows the division of the nucleus.

CLONE, CLONING. Two or more creatures produced by asexual reproduction are each said to be 'clones'; or the two (or more) collectively are said to form a 'clone'. 'Clone' is also used as a transitive or intransitive verb, as in 'the scientist clones the sheep' or 'potatoes clone naturally by stem tubers'.

CODON. The group of three bases in a DNA or RNA macromolecule that provides the code for each particular amino acid.

CONDENSED. The highly bunched up form adopted by chromosomes in the build-up to mitosis or meiosis, or after nuclear transfer into a MII oocyte.

COUPLET. The cytoplast and karyoplast before fusion.

CROSSING OVER. The process by which homologous chromosomes exchange genetic material during the first stage of meiosis.

CSF. An agent found especially in the oocyte which stabilises MPF. So long as CSF remains high, MPF remains high, and meiosis is arrested in metaphase II.

CULTURE. Many meanings, but applied here to the artificial cultivation of cells on culture mediums.

CYCLIN. A protein which joins with CDC2 to form the protein kinase enzyme that is MPF.

CYCLOHEXIMIDE. A chemical agent that can arrest the cell cycle.

CYTOCHALASIN B. A chemical agent that softens the cytoskeleton, facilitating microdissection and especially removal of the chromosomes (enucleation).

CYTOPLASM. The highly structured material of the cell that surrounds the nucleus.

CYTOPLASMIC RETICULUM, ENDOPLASMIC RETICULUM. The three-dimensional network of membranes within the cytoplasm which, among other things, holds the cell in shape; the cytoskeleton.

CYTOPLAST. In the present context, an enucleated cell (that is, cytoplasm) that is used to receive a new (donor) nucleus in the process of nuclear transfer.

CYTOSINE. One of the four bases (or nucleotides) that provide the only source of variability in DNA and RNA.

CYTOSKELETON. The network of fibres that move organelles within the cell and hold them in place.

DEOXYRIBONUCLEIC ACID. Alias DNA. The macromolecule of which genes are made. Each DNA macromolecule consists of two chains, coiled around each other to form a 'double helix'. Each chain is compounded of many 'building blocks', each of which contains a sugar (deoxyribose) and a 'nucleotide' or 'base', plus a phosphate radical. There are four kinds of base: cytosine (C), guanine (G), adenine (A) and thymine (T). The sequence of the bases provides the 'genetic code'.

DEUTEROSTOME. The group of animals that includes echinoderms (which include sea-urchins) and vertebrates.

DIFFERENTIATION. The process by which cells change in form and function as they develop and take on a specialist role.

DIPLOID. Pertaining to an organism, cell, or nucleus that contains two sets of chromosomes.

DNA. Short for deoxyribonucleic acid.

ELASTASE. An enzyme that helps to ensure the correct elasticity of the lungs.

ELECTROFUSION. The process by which the membranes of different cells are fused by applying an electric current.

EMBRYO DISC. Inner cell mass cells form themselves into an 'embryo disc' – really, the beginnings of the new animal – in advanced blastocysts.

EMBRYO STEM CELL (ES CELL). Loosely, a stem cell derived from an embryo. In the context of this book, more specifically applied to stem cells from particular strains of mice which, after culture, can be returned to the inner cell mass of an embryo and there demonstrate totipotency. They are able to develop into any of the tissues of the new host embryo.

ENUCLEATE. To remove a nucleus from a cell as carried out prior to nuclear transfer. However, Dolly and the other cloned sheep at Roslin were made by adding a new nucleus to the cytoplasm of MII oocytes. MII oocytes are arrested in mid meiosis and therefore do not strictly speaking contain nuclei, since at this stage the nuclear membrane has broken down, so that the chromosomes are free-floating. However, removal of these chromosomes is generally referred to as 'enucleation'.

ENZYME. A protein that acts as a biological catalyst.

ES CELL. See Embryo Stem Cell.

EUKARYOTE (EUCARYOTE). An organism whose cells contain distinct nuclei, which in turn contain the chromosomes. Animals and plants are eukaryotes. Eucaryote is the American spelling.

EXON. The part of a length of DNA that acts as a functional gene.

EXPRESS, EXPRESSION. Genes that are actually functioning in any one cell are said to be 'expressing' or 'expressed'.

FACTOR VIII. A protein that contributes to blood clotting.

FACTOR IX. Another protein that contributes to blood clotting.

FERTILISATION. The fusion of sperm with egg, whereby the sperm contributes genetic material to the egg.

FIBROBLAST. A flat and branched cell, typical of those found in skin; and also the form commonly adopted by animal cells in culture.

FOLLICLE CELL. One of the cells in the ovary that surrounds the oocyte and assists in its creation.

Go. 'G-nought' or 'G-zero'. The 'quiescent' stage of the cell which cells may enter when they are in G1.

GAMETE. A sex cell; sperm or egg.

GAMETOGENESIS. The production of gametes.

GAP 1 (G1). The first phase of interphase. The phase after mitosis (or meiosis) and before the S (synthesis) phase.

GAP 2 (G2). The third stage of interphase, after the S phase and before mitosis.

GASTRULA. An embryo that has already acquired two layers of cells.

GENE. The unit of heredity.

GENE KNOCKOUT. Removal of a functional gene.

GENE POOL. The total range of alleles in a breeding population.

GENE TARGETING. Insertion of a novel gene (transgene) into a precise point in the new host genome or the precise modification of an existing gene.

GENE TRANSFER. Taking a gene from one organism and putting it into another.

GENETIC DRIFT. The loss of allelic variants from a gene pool as the generations pass.

GENETIC ENGINEERING. The science and craft by which genes are transferred between organisms.

GENETIC FINGERPRINTING. The identification of particular organisms by studying the pattern of 'microsatellites' within their genome.

GENETICS. The study of genes and heredity.

GENOME. The total apportionment of genes within any one organism.

GENOME, GENOMIC ACTIVATION. The process in which the genes of a young embryo are first 'switched on' or 'expressed' and begin to synthesise proteins.

GENOMIC (DNA) CLONE. When two or more creatures have exactly the same DNA (nuclei) but have different cytoplasm, they are 'genomic' or 'DNA' clones. Dolly is a genomic clone of her 'clone mother'.

GENOMIC IMPRINTING. The process by which male or female parents put their own particular 'stamp' on some of the genes that they pass on to their offspring.

GENOMICS. The attempt to map and describe all the genes in various organisms.

GERM CELLS. The cells in the ovaries or testes that give rise to the gametes.

GERM LINE. The germ cells, which give rise to the gametes, develop separately from the body cells. Thus changes made in the body cells do not affect the germ cells, and are not passed on to the next generation. Changes made in the germ cells are passed on, however – and in turn manifest both in the body cells and in the germ cells of the following generation. Thus genes are in effect transmitted from the germ cells of one generation to the germ cells of the next – and this succession of germ cells is the 'germ line'.

GERMINAL LAYER. The layer of cells in the ovaries or testes that give rise to the gametes.

GLYCOPROTEIN. Material formed by a combination of protein and sugar, as in the protective *zona pellucida*.

GOAT. An acronym coined by Keith Campbell meaning 'Go Activation and Transfer'.

GROWTH FACTOR. All cells require 'growth factors' – commonly small proteins – to promote their growth.

GUANINE. One of the four bases found in DNA and RNA.

HAPLOID. An organism, cell, or nucleus containing only one set of chromosomes.

HATCH. The process by which a young creature emerges from an egg. Mammalian embryos are said to 'hatch' when they break out of the *zona pellucida*.

HeLa CELL. A line of 'immortal' cancerous cells originally derived (allegedly) from a woman called Helen Lane in the 1950s and now cultured in laboratories around the world.

HIGH TECHNOLOGY. As used in this book (and as originally defined by Colin Tudge), 'Technology that derives from science'.

HISTONE. A kind of protein, which forms the core of chromosomes around which DNA is entwined.

HOMOLOGUE. In a diploid organism, two equivalent chromosomes, each derived from the different parents.

ICSI (INTRA-CYTOPLASMIC SPERM INJECTION). A process of fertilising an egg by injecting sperm into it.

IMPLANT. In the context of this book, the process by which the embryo (blastocyst) becomes embedded in or attached to the wall of the uterus.

IN VITRO FERTILISATION (IVF). Fertilisation effected artificially by bringing egg and sperm together 'in glass' – in a test-tube or, more usually, in a Petri dish.

INNER CELL MASS (ICM). The central core of cells in a young mammalian embryo from which the embryo proper forms. It is surrounded by trophectoderm, or TE cells, which make the first contact with the wall of the uterus.

INSULIN. A proteinous hormone that regulates the concentration of sugar (glucose) in the blood. Lack of insulin (or lack of response to insulin) gives rise to diabetes.

INTERPHASE. The main part of the cycle of the cell, between bursts of division (mitosis or meiosis). It includes Gap 1, Gap 2 and S phase.

INTRON. A length of DNA that is found within the genes of eukaryotes, but appears to be non-functional (though it may have a function in regulating gene expression).

ISOGAMOUS. Some organisms, including some fungi, produce gametes of equal size, and are then said to be 'isogamous'.

JUNK DNA. Non-coding DNA. 'Junk' is an old expression. We should not assume that non-coding DNA has no function, and some of it clearly does (e.g. in regulating gene expression).

KARYOPLAST. A donor nucleus, with a greater or lesser amount of cytoplasm attached, which is transferred to new cytoplasm in nuclear transfer. A karyoplast may be an almost naked nucleus – or, as is typical at Roslin, may be a whole cell that is simply fused with the cytoplast.

LEUKAEMIA INHIBITION FACTOR (LIF). A biochemical agent that inhibits cell differentiation.

LICENSING FACTOR. An agent found within the cytoplasm of cells which, when allowed to enter the nucleus, stimulates replication of the DNA: i.e. stimulates entry into the S phase.

LIPOFECTION. Transfer of genes into cultured cells by enclosing the DNA in a fatty membrane, which facilitates entry through the cell membranes (which are also fatty).

MII OOCYTE. The egg at the stage at which it is extruded from the mammalian ovary in most mammals, ready to be fertilised. It is arrested in metaphase of the second stage of meiosis – hence 'MII'.

MACROMOLECULE. A large molecule formed by joining smaller molecules together. Protein and DNA are macromolecules.

MAGIC. An acronym, coined by Keith Campbell, meaning 'Metaphase-Arrested G1/G0 Accepting Cytoplast'.

MARKER. Various meanings, including a length of DNA (or a whole gene) that is easy to identify and indicates the presence of another gene; or a protein produced in cells, cultured or otherwise, which indicates that a particular gene is turned on and expressing.

MATERNAL TO ZYGOTE TRANSITION. See Genome Activation.

MEIOSIS. The process by which germ cells divide to form gametes. Meiosis takes place in two stages, meiosis I and meiosis II. In meiosis I, the homologous chromosomes exchange genetic material ('crossing over'). In meiosis II – 'reduction division' – the two resulting diploid cells with their recombined chromosomes divide further to form four haploid gametes.

MENDELIAN. Pertaining to the great Moravian geneticist, Gregor Mendel. Mendel showed how some characters (but not all) are inherited in simple ratios. Such ratios are called 'mendelian'.

METAPHASE. The phase of mitosis or meiosis between prophase and anaphase.

MICRODISSECTION. Dissection on a very small scale, e.g. of cells.

MICROSATELLITE. A small, repeated and highly variable stretch of DNA. Each individual has his or her own peculiar pattern of microsatellites; and this has given rise to the technique of 'genetic (or DNA) fingerprinting'.

MIDDLE PIECE. The central part of the sperm, rich in mitochondria.

MITOCHONDRION. A small structure – 'organelle' – within the cytoplasm of the cell which contains the enzymes required for respiration; i.e. for the provision of energy.

MITOSIS. The process by which somatic cells divide.

MORULA. The young embryo at the 'ball of cells' stage, before it becomes a blastocyst.

MPF (MATURATION PROMOTING FACTOR, MEIOSIS PROMOTING FACTOR, MITOSIS PROMOTING FACTOR). An enzyme that promotes mitosis and meiosis, and which induces nuclear envelope breakdown and chromosome condensation. It is an enzyme, a protein kinase, compounded from cyclin and cdc2.

MUTATION. The process by which alterations appear in genes. Most mutations have little obvious effect but some are deleterious and a very few bring improvements.

NUCLEAR ENVELOPE, NUCLEAR MEMBRANE. The membrane that surrounds the nucleus.

NUCLEAR ENVELOPE BREAKDOWN (NEBD). The breakdown of the nuclear envelope at the start of mitosis or meiosis.

NUCLEAR TRANSFER. Transfer of a karyoplast into cytoplasm from the different source.

NUCLEIC ACID. DNA or RNA.

NUCLEIN. The name first given to DNA by its discoverer, Johann Friedrich Miescher.

NUCLEOTIDE. Another name for 'base'. Nucleotides provide the only source of variability within DNA and RNA. DNA provides four different forms of nucleotide – thymine, cytosine, adenine and guanine. RNA contains uracil instead of thymine.

NUCLEUS. The body at the centre of a cell in eukaryotic organisms which contains the chromosomes. The nucleus stains dark with acid dyes because it contains the acid, DNA.

OOCYTE. The diploid egg, before meiosis is complete. Eggs are released from the mammalian ovary as MII oocytes; and MII oocytes are the favoured cytoplasts in the creation of embryos by nuclear transfer.

ORGANELLE. A small, specialist structure within the cytoplasm of a cell.

OVULATION. The release of eggs from the ovary.

PARTHENOGENESIS. Literally, 'virgin birth'. The development of embryos from unfertilised eggs.

PASSAGE. Pronounced 'pass-*arje*'. The period in which cultured cells continue to grow and multiply before they fill the dish or flask, when they have to be divided up and transferred to new containers.

PHARMING. A neologism describing the process by which thera-peutically valuable materials (usually proteins) are produced within genetically modified livestock; e.g. human AAT within the milk of Tracy.

PHENOTYPE. The physical (or behavioural) manifestation of a creature; as opposed to the 'genotype' which alludes to the genes that underpin the phenotype.

PLEIOTROPIC. Referring to a gene that has more than one effect on the phenotype of the organism that contains it.

PLURIPOTENT. Alluding to a cell that is able to divide to give rise to daughter cells of various different types.

POLAR BODY. A small entity containing surplus chromosomes discarded in the course of meiosis in the female.

POLYGENIC. A character that is 'determined' by more than one gene.

POLYMERASE CHAIN REACTION (PCR). Technique to multi-ply very small quantities of DNA quickly, to produce amounts large enough for analysis.

PREMATURE CHROMOSOME CONDENSATION (PCC). The tendency of chromosomes to condense as if entering mitosis when exposed to MPF, irrespective of the phase of the cell cycle.

PROKARYOTE (PROCARYOTE). An organism in which the genetic material is not contained within a nucleus (that is, within a nuclear membrane). In practice, prokaryotes (spelt procaryote in

America) are members of the two domains Bacteria and Archaea.

PROMETAPHASE. The phase of mitosis or meiosis between prophase and metaphase.

PROMOTER REGION. A stretch of DNA which produces a protein that prompts a gene to express or shut down.

PRONUCLEUS. The haploid nucleus of a sperm or egg.

PROPHASE. The first phase of mitosis or meiosis.

PROTEIN KINASE. An enzyme that transfers a phosphate radical to a protein.

PROTOPLASM. The old-fashioned name for 'cytoplasm', current before the structural complexity of cytoplasm was appreciated.

PSEUDOPREGNANCY. A physiological state in which a female produces the hormones appropriate to pregnancy, but is not pregnant.

QUIESCENT. In the context of this book: alluding to a state in which the cell is apparently 'shut down' (roughly analagous with 'hibernation' in a whole organism). Quiescent cells are said to be in Go: a variant of G1. Quiescence, however, does not merely imply a state of dormancy. Cells become 'quiescent' as a preliminary to differentiation; so the state seems to be associated with a re-programming of the genome.

RECOMBINANT DNA. Chemically synthesised DNA, DNA produced in a different species from the original gene (usually a bacterium) or DNA formed by joining DNA from two or more different sources. Since genes are made of DNA, recombinant DNA technology is the basis of genetic engineering.

RECONSTRUCTED EMBRYO. An embryo made by nuclear transfer; introducing a karyoplast into a cytoplast that generally comes from a different individual.

REPLICATION. Exact multiplication of an organism, a cell, DNA, or any other entity.

RETROVIRUS. A virus that is able to induce its host cell to use the virus's own RNA as a model from which to construct DNA – thus reversing the 'dogma' which says that 'DNA makes RNA'.

RIBONUCLEIC ACID (RNA). Various forms of nucleic acid that act as intermediaries between DNA and protein. We might say that DNA is the administrator, and RNAs are the executives. RNA has the same kind of structure as DNA except that its macromolecules are single stranded rather than double stranded, and it contains the base uracil instead of thymine.

RIBOSOME. Organelles within the cytoplasm where various RNAs cooperate to join amino acids together to form proteins.

RNA. See Ribonucleic Acid.

ROSNI (ROUND SPERMATIC NUCLEUS INJECTION). Injection of spermatids (haploid male gametes that are not yet mature) into the cytoplasm of an oocyte to effect fertilisation artificially.

S PHASE. The phase of the cell cycle in which DNA is replicated.

SENDAI VIRUS. A virus used to fuse different cells. In cloning, now largely superseded by electrofusion.

SEXUAL DIMORPHISM. The condition in which the two different sexes look different.

SOMATIC. Relating to the body. As applied to cells: body cells as opposed to germ cells.

SPERMATOCYTE, SPERMATID, SPERMATOZOON. The spermatozoon (plural spermatozoa) are the mature haploid male gametes, with a head, middle piece and tail. Spermatids are the haploid precursors of the spermatozoa, which have not yet developed distinct heads and tails. Spermatocytes are the diploid male germ cells, which give rise to spermatids by meiosis.

SPINDLE. The structure that forms in the cytoplasm of the dividing cell and drags the different sets of chromosomes apart during mitosis and meiosis.

STEM CELL. A cell that may look relatively undifferentiated, but which continues to divide to produce differentiated offspring. Some stem cells are 'pluripotent', like those that give rise to red blood cells and the several different kinds of white blood cell. Some are totipotent.

SUPEROVULATION. Stimulation of an ovary to release more than the usual number of oocytes at one time.

SURROGACY, SURROGATE. Literally, a deputy, or substitute. Hence a surrogate mother is one who gestates, and gives birth to, an embryo that is introduced into her.

TELOMERE. A section at the end of a chromosome which, in effect, prevents it from 'fraying'. Every time the chromosome divides some of the telomere is lost, so that cells in old individuals have shorter telomeres than those in young individuals. Dolly's telomeres are shorter than is usual in sheep of her age.

TELOPHASE. The final phase of mitosis or meiosis.

TETRAPLOID. An organism, cell, or nucleus containing four sets of chromosomes.

THYMINE. One of the bases in DNA. Replaced in RNA by uracil.

TNT4 CELL. An acronym coined by Jim McWhir at Roslin meaning 'Totipotent for Nuclear Transfer'. The term applied to cultured cells which retained sufficient totipotency to give rise to entire, healthy lambs when transferred into suitable cytoplasts.

TOTIPOTENCY. The ability of a cell to give rise to descendant cells that may differentiate to form any of the kinds of tissues typical of the organism.

TRANSCRIPTION. The synthesis of a molecule of RNA that corresponds to a particular stretch of DNA (i.e. to a gene).

TRANSFECTION. The act of 'infecting' a cell with DNA to effect gene transfer.

TRANSFORM. To alter a cell or organism by adding DNA (one or more genes) to its genome.

TRANSGENE, TRANSGENIC. A gene transferred from one organism into another.

TROPHECTODERM (TE). The cells in the outer part of the young mammalian embryo which make the first contact with the wall of the uterus.

UNIVERSAL RECIPIENT. A cytoplast with a low MPF content, able to play host to a donor nucleus at any stage of its cell cycle.

URACIL. A base found in RNA but not in DNA (which contains thymine instead).

VITELLINE SPACE. The space between the oocyte (or young embryo, as it becomes) and the *zona pellucida*.

XENOTRANSPLANT. An organ or tissue transplanted from one species of animal into another.

ZONA PELLUCIDA. The glycoprotein 'shell' that surrounds the oocyte and young embryo before implantation.

ZYGOTE. The one-cell embryo formed by the fusion of sperm and egg.

INDEX

AAT (alpha-1-antitrypsin) 31–2, 35–6, 46–7, 50–52, 234, 262, 277
agar 149–50, 153, 174, 212
alleles 41, 315–16
allografts 278–9, 281
Altmann, Richard 36
amino acids 38–9, 72
anaphase 108, 110, 112
Andrews, Lori 247, 297
aneuploid cells 189, 215, 235
Animal Breeding Research Organisation (ABRO) 7–8, 30–31, 50, 55, 171–2; see also Roslin Institute
Ansell, Ray 214–15
armadillos 66
artificial insemination (AI)
 of animals 155–6, 271–2, 293
 of humans 293, 295–8, 322
asexual reproduction 60–66
Avery, Oswald 37, 91

Barnes, Frank 155
Berg, Paul 43
beta-lactoglobulin promoter 51, 259
binary fission 60–61
biological control 17, 24, 267, 287, 288–90
Biology of Reproduction (journal) 175, 199
Biotechnology (journal) 51
Biotechnology and Biological Sciences Research Council (BBSRC) 4
Bishop, John 7, 30–31, 50
Bishun, Nutan 181
bivalents 111–12
Bladerunner 310–11
blastocysts 79–80, 116, 119, 168
Blow, Julian 183

Bohr, Niels 331
Bonnie 2, 148, 319
Bowran, Harry 237
Boys from Brazil 129, 306
Bracken, John 209, 213, 221, 225, 232, 237, 239
Brenner, Sydney 38
Briggs, Robert 30, 89–90, 93–4, 96, 288
Brinster, Ralph 295
Brockman, John 2
Bromhall, Derek 129, 132
Brown, Louise Joy 293
Bulfield, Grahame 171, 178, 245

calcium concentration 197–8, 219–20
calf embryo, freezing of 26–7, 127
callus cells 69–70
cancer 22, 27, 55, 90, 182, 190, 280
Capron, Alexander Morgan 305
cdc2 196
Cecil, Cedric and Cyril 2, 17, 22, 74, 238, 242, 304, 307
Cell (journal) 131, 135–6
cell culture 164–6, 235–6
cell cycle, the 28–9, 104, 142, 176–93, 203
 control of 193–6
 link with cloning 228
cell cycle experiments 217–27
cells, division of 102–8, 119–21, 328
centromeres 106–8
chimeras 54–5, 70, 150–51, 163, 173, 258, 283
chromatids 106, 108, 111–12, 187, 192
chromatin 72, 105, 195
chromosomes 40–42, 64, 101–12,